PREVENTION OF
STRUCTURAL FAILURES

— The Role of NDT, Fracture Mechanics
and Failure Analysis

PREVENTION OF
STRUCTURAL FAILURES

— The Role of NDT, Fracture Mechanics
and Failure Analysis

Proceedings of Two Annual Forums
19-22 June 1977, and 14-16 June 1976
Tarpon Springs, Florida

Co-sponsored by
Materials Testing and Quality Control Division
American Society for Metals
and
Technical Council
American Society for Nondestructive Testing

Materials/Metalworking Technology Series
AMERICAN SOCIETY FOR METALS
Metals Park, Ohio 44073

Library of Congress Cataloging in Publication Data

Main entry under title:

Prevention of structural failures.

   (Materials/metalworking technology series)
   Includes bibliographical references.
   1. Structural stability—Congresses.
2. Non-destructive testing—Congresses. 3. Fracture
mechanics—Congresses. I. American Society for
Metals. Materials Testing and Quality Control
Division. II. American Society for Nondestructive
Testing. Technical Council. III. Series.
TA656.P73   624'.171   78-15388
ISBN 0-87170-006-9

PRINTED IN THE UNITED STATES OF AMERICA

FOREWORD

The Materials Testing and Quality Control Division of
the American Society for Metals (ASM) and the Technical
Council of the American Society for Nondestructive Testing
(ASNT) jointly sponsored the Fifth Annual Forum on
Prevention of Structural Failures in the Transportation
Industry. The Forum was held on June 20-22, 1977, at the
Innisbrook Resort Hotel in Tarpon Springs, Florida, and was
subtitled "The Role of Fracture Mechanics and NDT in Failure
Control." This publication comprises selected papers
presented at the Forum.

Also included in this publication are selected papers
from the Fourth Annual Forum entitled "Prevention of Failure
Through Nondestructive Inspection." This Forum was held
June 14-16 1976, also at the Innisbrook Resort Hotel.

INTRODUCTION

WILLIAM H. LEWIS*

The 1977 Fifth Annual Forum on the "Prevention of Structural Failures in the Transportation Industry" was a continuation of the meetings on the fracture mechanics/nondestructive testing interrelationship which were begun by the Materials Testing and Quality Control Division of ASM in 1973. These meetings have had all the ingredients for turning out highly successful forums: expert, knowledgable speakers; interesting and timely topics covering a broad range of the transportation world; responsive attendees; lively discussions; well-run sessions; and most pleasant surroundings. The 1976 and 1977 Forums, from which these papers were selected, were certainly no exception.

The main topic of these Forums is highly important in the design of new systems and throughout the service/maintenance life of new and older systems. As made abundantly clear in the papers selected for this publication, case histories continually underline the need for designing for damage tolerance and inspecting for that damage to prevent failures during the expected life span of the system. As far back as the 15th century, the first requirements for Damage Tolerant Design apparently were laid out in the notebooks of Leonardo da Vinci, in which he discussed the

---

*Mr. Lewis, NDI Program Manager at Lockheed-Georgia Company, is Chairman of the Technical Council of the American Society for Nondestructive Testing and served as Program Chairman for the Fifth Annual Forum on Prevention of Structural Failures in the Transportation Industry.

design of flying machines:

> "In constructing wings one should make one cord to
> bear the strain and a looser one in the same
> position so that if the one breakes under strain
> the other is in position to serve the same function."

Perhaps he should have gone on to indicate that
"inspections should be performed to determine if the first
cord is broken, and if the safety margin no longer remains,
further operation should cease until repairs are made."

No structure can reasonably be expected to last
forever. Even the fabled London Bridge eventually had to
give way to the concerns of degradation and the structure's
ability to withstand modern load demands. Could the
Lafayette Street Bridge catastrophe have been avoided if
better life prediction methods were applied to remove it
from service at the end of a predictable safe-life span,
using fracture mechanics and nondestructive testing considerations at the outset? The life-limiting potential of
undetected flaws that grow under the influence of loads and
environments can affect any man-made structure. Hence, a
major aim of the fracture mechanics/nondestructive testing
(FM/NDT) marriage is to deal with the realistic possibility
that flaws have escaped initial and in-service inspections,
and to compensate for these undetected flaws with reduced
design stress levels or predicted spans of safe-life
operation.

Effective application of FM principles requires quantified NDT data: discrete, specific information about defect
shape, size, orientation, and location. Such quantified NDT
data are not easy to come by with present techniques and
methods. Perhaps forums such as this one can help to usher
in a new era of NDT technology that can provide the needed
data. In recent years, programs have been conducted to
measure how effective nondestructive inspection is at finding crack-like defects over a broad range of sizes and
material configurations. While the results will not bring
cheers from the structural designer, they do provide quantified estimates of what percentage of defects can be missed
for specified defect sizes, NDT techniques, and test conditions. Certainly, a gap in the technology is being filled.

The impact of nondestructive inspection is felt in
operations concerned with both aging structures and new
systems. The current drive to maintain older systems,

rather than build new ones, to extend the service life of aging structures (aircraft and bridges, for example) beyond the useful period initially planned, places a much greater burden on NDT. Inspection intervals must be planned which are compatible with flaw growth characteristics in the system. Fracture mechanics will be increasingly called upon to define damage-tolerance limits for given defect sizes and to estimate projected flaw sizes due to growth under specific service conditions. The purposes are combined into one mutual goal: to help assure an extended period of safe life at minimum maintenance expense. It is believed that the NDT and fracture mechanics disciplines will have tremendous roles in the future conservation of our resources and their use in hardware planned for longer safe life with minimum maintenance.

Almost daily we read of incidents in which this technology could have been of immeasurable value if it had been applied in time: an oil tanker breaks up in heavy seas or runs aground and the oil spill becomes a double-edged sword - a valuable resource is wasted and a beautiful seacoast is contaminated; a bridge collapses; a dam breaks and the flood sweeps away lives, homes, and property; an airplane loses a door and a disastrous crash ensues; a rail splits and a train loaded with a cargo of poisonous gas is piled up like cordwood. Operational errors aside, these examples and others addressed in this Forum point out the serious purpose and the challenge of our profession. While the pleasant, subtropical environment we enjoyed during these meetings may belie the seriousness of our concerns, the efforts of all who were involved will eventually reduce or eliminate the costs in lives, property, wealth and resources resulting from such disasters.

Both the 1976 and 1977 Forums held at the Innisbrook Resort Hotel in Tarpon Springs, Florida were co-sponsored by the Technical Council of the American Society for Nondestructive Testing and the Materials Testing and Quality Control Division of the American Society for Metals, with both organizations sharing technical program responsibilities. With regard to the 1977 Forum, I would like to recognize the efforts of Dr. Paul F. Packman, Southern Methodist University, and Mr. Nathan G. Tupper, U. S. Air Force Materials Laboratory, for serving as Session Chairmen; and Mr. Thomas D. Cooper, U. S. Air Force Materials Laboratory, Mr. John R. Cope, Newport News Shipbuilding and Dry Dock Co., Mr. Charles Galambos and Mr. Michael Lauriente, of the U. S.

Department of Transportation, who were Panel Moderators
during the lively discussion period.

I also wish to thank Mr. Charles H. Scheffey, Director,
Office of Research, Federal Highway Administration, for
serving as my co-chairman for the Fifth Annual Forum and
providing Department of Transportation support and encouragement for the Technical Program.

Finally, I especially want to thank my colleague,
Mr. William M. Pless of the Lockheed-Georgia Company, who
worked with the authors in arranging final program details
and finalizing this document for publication.

While we feel justly proud of these successful Forums,
we welcome your suggestions and participation to help
produce even better meetings in coming years.

WHL

April 26, 1978

Note: The 1976 Forum Co-Chairmen were Dr. Paul F.
Packman, Southern Methodist University, and Mr. Don Conn,
Armco Steel Corporation.   The Program Chairmen were
Mr. Thomas D. Cooper, U. S. Air Force Materials Laboratory,
and Dr. B. G. W. Yee, General Dynamics Corporation.

CONTENTS

Fifth Annual Forum, 1977:
Prevention of Failure in the Transportation Industry

Fourth Annual Forum, 1976:
Prevention of Failure Through Nondestructive Inspection

PREVENTION OF
STRUCTURAL FAILURES

— The Role of NDT, Fracture Mechanics
and Failure Analysis

ASM/ASNT Fifth Annual Forum

19-22 June 1977

PREVENTION OF FAILURE IN THE

TRANSPORTATION INDUSTRY

THE FAILURE OF THE LAFAYETTE STREET BRIDGE
- INFLUENCE OF DESIGN DETAILS

John W. Fisher[a], Alan W. Pense[b] and Richard Roberts[c]

[a]Professor of Civil Engineering and Associate Director, Fritz
Engineering Laboratory, Lehigh University, Bethlehem, Pa.
[b]Professor of Metallurgy and Materials Science
[c]Professor of Mechanical Engineering and Mechanics

INTRODUCTION

On May 7, 1975, one of the main girders of the Lafayette
Street Bridge over the Mississippi River in St. Paul,
Minnesota, was discovered to be cracked.

The crack was discovered in the east girder of the
southbound structure in Span 10. Span 10 is the 362 ft.
(110 m) main span of a three-span structure with side spans
of 270 ft. (82.3 m) (Span 9) and 250 ft-6 in. (76.4 m)
(Span 11). Forty ft. (12.2 m) cantilevers project into
Spans 8 and 12 as shown schematically in Fig. 1. The web and
flange were fabricated from ASTM A441 steel.

Figure 2 shows the crack that developed in the girder
web and the lateral bracing gusset plate. The crack occurred
about 118 ft-8 in. (36.2 m) from Pier 10. The crack had propagated in the web to within 7½ in. (190 mm) of the top
flange when discovered on May 7, 1975. The entire bottom
flange was fractured.

The structure was opened to traffic on November 13,
1968. The southbound lanes were closed from May 20, 1974
through October 25, 1974 for repairs to the deck and over-
lays. The estimated average daily truck traffic crossing the
structure was 1500 vehicles during the period November 1968
through May 1975. Hence, approximately 3,300,000 trucks had
crossed the span at the time the fracture was discovered.

A detailed study of the fractured portions of the web,
flange and gusset plate was made using material that was

3

removed from the cracked girder. Figure 3 shows a schematic
of the cross section and indicates the materials that were
removed at the cracked cross section. They included the entire fracture surfaces of flange, both fracture surfaces of
the web extending well above the gusset plate, portions of
the gusset and transverse stiffener.

PHYSICAL PROPERTIES OF FLANGE, WEB AND GUSSET PLATE

(a) Physical and Chemical Properties

The material in the flange, web and gusset plates was
evaluated to provide information on the chemical and physical
properties of the plates. This included: an evaluation of
the chemistry of the plates; Charpy V-notch tests of the
plates; physical tests; and K_{IC} tests of the web plate.

The physical properties of the web, flange and gusset
plate are listed in Table 1(a).

TABLE 1   PHYSICAL PROPERTIES

(a) Mechanical Properties

	Thickness in. (mm)	Yield Stress ksi (MPa)	Tensile Strength ksi (MPa)	Percent Elongation	Percent Red. in Area
Flange (A441)	2½ (64)	46.0 (317)	76.2 (525)	32.8	65.8
Web (A441)	½ (13)	53.7 (370)	81.8 (564)	30.0	68.0
Gusset (A36)	½ (13)	37.9 (261)	67.0 (462)	36.4	67.1

(b) Chemical Properties

	C	Mn	P	S	Si	Cu	Va
Flange	0.19	1.18	0.046	0.030	0.23	0.25	0.046
Web	0.23	1.17	0.024	0.044	0.14	0.18	0.032
Gusset	0.25	0.87	0.038	0.040	0.06	0.016	0.004

The physical properties of the flange and web plate are
in good agreement with ASTM Specification A441. Both pieces
exhibited good elongation and reduction in area characteristics. The gusset plate was in good agreement with ASTM
Specification A36.

The chemical composition of the flange and web plates
were obtained by check analysis and are listed in Table 1(b).
The chemical composition is in reasonable agreement with the
check analysis of the ASTM Specifications*. The flange and
web plate show correspondence with the requirements of A441,
and the gusset plate with A36.

(b) Fracture Toughness

The fracture toughness of the web, flange and gusset
plates were determined by making longitudinal V-notch Charpy
tests and K_{IC} tests.

Transition temperatues curves of the web and flange A441
steel showed that the 15 ft-lb. (20.4 J) transition temperature is about 50° F (10° C) for the flange and 15° F (-10° C)
for the web. The web and flange exhibited nearly identical
behavior.

The flange plate meets the Charpy V-notch requirements
of the 1974 AASHTO Specification for Temperature Zone Designation 1. [Minimum service temperature to -0° F (-18° C)].
The web plate meets the requirements for Temperature Zone 2
[-30° F (-34° C)]. The gusset plate material provided
15 ft-lb. (20.4 J) at 55° F (13° C) and met the requirements
for Temperature Zone 1.

The fracture toughness K_{IC} of the web and flange
steel plates can be estimated from Charpy V-notch impact
values and the empirical correlations published in ASTM
STP 466 and companion reports (2). Barsom, in Ref. 2, has
suggested that K_{IC} be estimated for intermediate strain rates
by using a temperature shift $T = 215 - 1.5 \sigma_y$ between dynamic
and static values (°F and σ_y = ksi). The fracture toughness
K_I was suggested to be $K_I^2 = 5E$ (CVN).

* ASTM A441-63T, ASTM A36-62T

Since an intermediate strain rate provides a temperature shift of about 75% from the dynamic curve, a minimum service temperature of -30° F (-34° C) would correspond to a test temperature shift of 0.75 [2.5 - 1.5 (54)] = 100° F (55° C). A 50% shift would yield about 67° F (37° C). This suggests that $K_{IC(Inter)}$ should be estimated from a Charpy V-notch value of 30 - 35 ft-lbs. (40 - 48 J). This yields an estimated $K_{IC(Inter)}$ value ∿ 70 ksi√in. (77 MPa/m$^{3/2}$).

To provide a more accurate assessment of fracture resistance four compact tension, K_{IC} tests (3) were run on material from the web plate. One test, Specimen No. 1 was run at 0° F (-18° C) and an intermediate strain rate (a loading time of about one second from zero to maximum load).

The remaining three specimens were run at static loading rates, 100 seconds from zero to maximum load, and temperatures of -25° F (-32° C), -120° F (84° C) and -150° F (-101° C). The results of these tests are given in Table 2.

TABLE 2 K_{IC} FRACTURE TESTS

Spec. No.	Test Temp. F (C)	Crack Depth a_i in. (mm)	a/w	Failure Load P_Q lbs. (N)	K_Q ksi (MPa √in. /m$^{3/2}$)
1	0° (-18°)	1.20 (30.5)	0.480	5,875 (26133)	68.0 (74.8)
2	-25° (-32°)	1.20 (30.5)	0.480	6,850 (30470)	79.3 (87.2)
3	-120° (-84°)	1.23 (31.2)	0.491	4,300 (19127)	51.3 (56.4)
4	-150°(-101°)	1.24 (31.5)	0.496	4,850 (21574)	58.6 (64.5)

All K results given in Table 2 should be considered as K_Q values as the specimens did not meet the thickness requirement of ASTM Specification E399 (3). Furthermore, Specimens 1 and 2 did not meet the E399 linearity requirement for load displacement records while Specimens 3 and 4 did. It is interesting to note the agreement between the measured K_Q value of 68 ksi√in. (74.8 MPa/m$^{3/2}$) for Specimen 1 and the estimated value of 70 ksi√in. (77 MPa/m$^{3/2}$) discussed earlier in this section.

FRACTOGRAPHIC EXAMINATION OF FRACTURED GIRDER

Figure 4a shows a closeup of the fracture surface of the gusset-transverse stiffener connection and the adjacent web. The mating face of the gusset and web is shown in Fig. 4b.

The fracture surfaces indicate that the transverse and longitudinal single bevel groove welds that connect the gusset plate to the transverse stiffener and web did not penetrate to the backup bar on the root of the weld preparation. As a result, significant amounts of lack of fusion existed. This was particularly acute in the corner of the web-stiffener junction as is visually apparent in Fig. 4. Close examination indicated that the lack of fusion varied from about 0.15 in. (3.8 mm) near the edge of the transverse stiffener up to 0.38 in. (9.65 mm) near the web. Figure 5 shows the fracture surface of the web-flange junction. The stiffener located on the outside girder face was burned off during removal of the section from the structure.

Visual examination of the fracture surface indicated that the fatigue crack growth had originated in the weld between the gusset plate and the transverse stiffener as a result of the large lack of fusion discontinuity in this location. The fracture surface of the web indicated that a cleavage (brittle) fracture had occurred as a result of the crack propagating into the web through the gusset-stiffener weld. The gusset-stiffener weld intersected the transverse stiffener welds that attached the stiffener to the web. As can be seen in Fig. 5, cleavage fracture extended down the web and into the flange.

Visual examination by Minnesota Department of Highways personnel shortly after the fracture was discovered indicated that a shiny fracture surface could be seen in the web from a point 4 to 6 in. (100 to 150 mm) above the gusset to the end of the crack near the top flange. The remaining portion of the web and the entire flange were coated by oxides from their exposure to the environmental conditions. By May 18, 1975, the entire fracture surface was coated by oxides and the fracture surface had a near uniform appearance.

Surface replicas for fractographic analysis were made at several locations on the fracture surface. These replicas were used to examine the fracture with the transmission electron microscope.

The replicas were taken at five locations. These included two locations at the gusset-stiffener connection near the web where fatigue crack growth was apparent; in the flange near the web-flange connection to ascertain whether or not the crack had arrested in the flange and experienced fatigue crack growth; in the web about 4 in. (100 mm) above the gusset where the crack surface was reported to be free of oxides; and in the gusset-stiffener weld about 5 in. (125 mm) from the web.

Figures 6a and b show fatigue crack growth striations in the gusset plate-transverse stiffener weld in the girder web. These striations, a lineation seen on the photograph, mark the progression of a fatigue crack through the steel structure and are a sure sign of fatigue fracture.

The striations near the large lack of fusion condition adjacent to the girder web were moving on an angle into the girder web. The crack growth rate was observed to be about 1.10×10^{-6} in./cycle (2.79×10^{-5} mm/cycle). The fatigue crack propagated into the girder web from the transverse stiffener-gusset weld connection. This region was observed to exhibit considerable abrasion which is believed to be due to the blast cleaning of the girder in the crack vicinity after the fracture. Nevertheless, fatigue crack growth striations were still occasionally present as is illustrated in Fig. 6a. The striations shown in Fig. 6b were located between 0.3 and 0.4 (7.6 to 10 mm) from the gusset-web face.

A substantial amount of abrasion was observed in the web above the gusset near the shiny-rusted boundary of the web observed by the Minnesota Highway personnel. Crack growth striations were detected over a small region and indicated that at least 10,000 stress cycles had occurred before cleavage fracture was again evident.

An examination was also made of the fracture surface in the flange near the flange-web connection. None of the replicas showed any fatigue crack growth. The fracture surface showed a large number of cleavage facets which indicated that the flange fracture is essentially brittle in nature.

FRACTOGRAPHIC EXAMINATION OF PLUG

A second crack was detected after the fracture was repaired one floor beam north of the fracture 94 ft-6 in. (28.8 m) from Pier 10. This crack was about 1 in. (25.4 mm) long on the web surface. A plug was removed from this area and also examined. Figure 7a shows a photograph of part of the

plug after saw cuts were made into the web and the crack
surfaces exposed. The plug included a portion of the gusset
plate and the web-to-gusset groove weld. After a circular
cut was made into the web, gusset and transverse stiffener,
a torch was used to cut the resulting plug free. The torch
marks are readily apparent on the gusset and part of the
weld. Although this destroyed the crack surface on the gusset, it did not damage the cracks that were exposed in the
web plate.

Figure 7b shows the exposed crack faces. This revealed
three crack surfaces in the web. These surfaces can be dif-
ferentiated by observing the light areas which mark their
boundaries. These light areas represent metal that was
fractured as the cracks were exposed. Figure 8 shows a
schematic diagram of the fracture surfaces and identifies
the three crack areas with Roman numerals I, II, and III.

Area I shows that the crack propagated into the web
from the transverse groove weld connecting the gusset to the
stiffener as was observed at the fractured cross section.
Area II was completely inside the web and could not be seen
from the web surfaces as thin ligaments along the crack
edges were still intact. Area III was similar to Area II.
However, this portion of the crack had penetrated the web
surface adjacent to the gusset plate and was detected by
Minnesota Department of Highways personnel after the jack
loads were removed from the fractured girder after the repair. The crack had not been detected prior to this operation.

To assist in evaluating the crack surface, replicas
were made at several locations on the fracture surfaces.
These replicas were used to examine the fracture surface
with the transmission electron microscope. The surfaces
were also examined with the scanning electron microscope.

Areas II and III were very clean and had very little
oxide formation on the surfaces which would be expected since
these cracks were inside the web and protected from the environment.

The examination of the fracture surfaces revealed that
all replicas in Area I had fatigue crack growth striations.
The crack growth rate observed was about 1.3×10^{-6} in./
cycle (3.3×10^{-5} mm/cycle). Hence, all of crack Area I was
created by fatigue crack growth.

The examination of crack Area II showed that cleavage (brittle) fracture existed throughout the area except for the crack tip at the bottom of Fig. 8. Crack growth striations were observed near the lower 0.075 in. (1.9 mm) at the crack tip.

ANALYSIS OF CRACK GROWTH

These facts are known from inspection and the observations of Minnesota Department of Highways personnel.

1. On May 18, 1975, the entire fracture surface was coated by oxides and the fracture surfaces had a near uniform appearance.

2. On May 7, 1975, one of the main girders was discovered to be cracked (see Fig. 2) and a 7½ in. (190 mm) difference level of medians of the two adjacent structures was observed near the fracture. The web fracture surface indicated shiny metal 4 to 6 in. (100 to 150 mm) above the gusset to the end of the crack near the top flange. The web below the gusset and the entire web were completely coated by oxides.

3. On March 20, 1975, a 2½ in. (63.5 mm) difference in the level of medians was measured of the two adjacent structures.

4. Prior to March 20, 1975, the temperature at St. Paul was recorded at

These observations when coupled with the material characteristics and the fractographic examinations can be used to help model the crack growth and explain the sequence of fracture.

(a) At Failed Section

The fracture surfaces showed clearly that several stages of crack growth had occurred in the Lafayette Street Bridge. Both fatigue and brittle fracture modes were observed. Figure 9 shows schematically the apparent crack stages in the gusset-transverse stiffener weld and gusset plate, the web, and the flange. Stage I corresponded to the initial crack condition which resulted from the lack of fusion in the gusset-transverse stiffener connection. This resulted in a long semielliptical shaped crack which was about 0.38 in. (9.65 mm) deep.

Under normal truck traffic, fatigue crack growth could
occur from this large initial crack condition and is shown
as Stage II. Fatigue crack growth occurred all along the
gusset-transverse stiffener weld. Near the web, the crack
was simultaneously drawn toward and into the web surface by
the stress concentration provided by the longitudinal groove
weld connecting the gusset to the web and the fact that the
transverse weld intersected the stiffener weld and web. The
striations shown in Fig. 6b confirmed the existence of fatigue crack growth in the web up to about 0.35 in. (8.9 mm)
of penetration.

An estimate of the time required to propagate the initial crack through the gusset plate thickness and into the
web was made using the stress intensity range defined in
Ref. 6 for crack growth at a weld toe. This is defined as

$$K = K_T \left[1 - 3.215 \frac{a}{t} + 7.897 \frac{a^2}{t} - 9.288 \frac{a^3}{t} + 4.086 \frac{a^4}{t} \right]$$

$$x\ \sigma \left[\frac{1 + 0.12\ (1 - a/b)}{\Phi_o} \right] \sqrt{\pi a}\ \sqrt{\sec \frac{\pi a}{2t}}$$

(1)

where K_T is the stress concentration = 2.64

  a is the crack depth = 0.38 in. (9.65 mm) for the
   initial crack size
  t is the gusset plate or web thickness = 0.5 in.
   (12.7 mm).

The initial crack depth was taken as 0.38 in. (9.65 mm) and
the ratio of b/a about 6 for this stage of crack growth. The
root-mean-square stress range was approximated from the gross
vehicle weight distribution given in Ref. 7. A standard HS20
truck generated a moment range in the main girder which varied from +3814 ft-kips (5183 KN-m) to -1622 ft-kips (2204
KN-m). This results in a stress range of 4.68 ksi (32.3 MPa)
in the girder flange and 4.13 ksi (28.5 MPa) at the gusset-
web connection. Since the girder is likely to act composite-
ly with the slab, the strain gradient may be less which will
not decrease the stress at the gusset as much as implied by
non-composite action. A root-mean-square stress range,
SrRMS = 2 ksi (13.8 MPa) was estimated for the gusset connection assuming that the average daily truck traffic was compatible with the 1970 FHWA Nationwide Loadometer Survey (7).

The cyclic life was estimated from the crack growth
relationship given in Ref. 6. This results in:

$$\frac{da}{dN} = 2 \times 10^{-10} \, \Delta K^3 \tag{2}$$

The life was estimated by integrating the crack growth relationship as the crack penetrated the gusset and web plates. When Eq. 1 was used to estimate the cyclic stress intensity factor, it yielded a fatigue life of about 3×10^6 cycles of random loading. An estimate was also made considering a simple edge crack in the gusset plate and about the same fatigue life resulted. Since the crack was observed to simultaneously propagate into the web, an equivalent center cracked plate was used to estimate the web penetration. This indicated that a fatigue crack would grow about 3/8 in. (9.5 mm) after 3×10^6 cycles.

The average daily truck traffic crossing the bridge is about 1500 ADTT. Hence, the gusset plate would be predicted to have the fatigue crack propagated through the plate thickness by the end of 1974 and the web would have been nearly penetrated.

At this point a through crack would exist in the gusset and the web would be penetrated about 3/8 in. (8.9 mm). Subsequent crack growth would occur in the girder web and toward the edges of the transverse stiffener. The eighty day interval between the end of 1974 and the first possibility of trouble on March 20, 1975 when a $2\frac{1}{2}$ in. (63.5 mm) difference in level of medians was measured would provide ample time for further crack propagation in the web and gusset. About 120,000 cycles of random stress would occur during this period.

A web crack in the region between the end of the gusset weld to the web and the transverse stiffener weld would exist in a stress field equal to the web yield point of 54 ksi (373 MPa). The close proximity of these two weld toes and the length of longitudinal weld insures a high residual stress condition. The stress concentration condition is somewhat more severe at the end of the long longitudinal groove weld than at the gusset transverse stiffener weld. An estimate of the critical crack size in the web can be made by equating Eq. 1 to the fracture toughness K_{IC} for an intermediate loading rate. An estimated value of 70 ksi/in. (77 MPa/m$^{3/2}$) appears reasonable for the web as the brittle fracture likely occurred either Februrary 9, 1975 when the temperature reached -22°F (-30°C) or on March 13 when it was -8°F (-22°C). With an assumed K_{IC} value of 70 ksi/in. (77 MPa/m$^{3/2}$), a critical crack size of about 0.35 in. (89 mm) results for the web when Eq. 1 is equated to K_{IC}. If the

web crack is considered as a semicircular-shaped surface
crack, a comparable crack radius yields about the same level
of stress intensity at the yield stress level.

This estimated web crack for brittle fracture in the
web is in agreement with the fracture surface shown in Fig.
4. The rapid fracture chevrons in the web on each size of
the gusset plate point toward the intersection of the transverse gusset weld and the web.

The formation of the brittle fracture in the web corresponds to Stage III of the crack formation. No evidence of
crack arrest was detected in the flange. This seems reasonable in view of the Charpy V-notch test data. The fracture
toughness of the web and flange were comparable with the
flange showing slightly less fracture toughness.

The brittle fracture arrested in the web about 6 in.
(153 mm) above the gusset plate. At this point there are no
significant residual tensile stresses. Furthermore, the
dead load stresses were relatively small. In addition the
gusset plate was bridging the crack and prevented the crack
from opening excessively.

Following the formation of the initial brittle fracture,
subsequent fatigue crack growth occurred in the web and in
the gusset plate. This growth likely occurred during the
interval of time between February 9 and April 28. It
appears that the gusset plate provided a number of arrest
conditions during this period of time.

Final fracture of the web occurred as further fatigue
crack growth and yielding of the gusset plate occurred
which permitted the crack to open. This eventually led to
the final girder fracture.

(b) At Section Which Was Plugged

The cracked surfaces of the plug removed from the girder exhibited comparable behavior during the initial stage
of crack growth.

The examination of the cracked surfaces indicates that
the following sequence of crack growth occurred. The fatigue crack propagated into the web in Area I (see Fig. 8)
at the same time the crack was propagating through the transverse groove weld in the gusset-transverse stiffener connection. This is directly compatible with Stage II of crack
growth which was experienced at the major fracture.

14 / J. Fisher, A. Pense, R. Roberts

At this stage of crack growth a limited pop-in brittle crack occurred which is Crack Area II in Fig. 8. The fatigue crack growth striations at the bottom of Crack Area II indicate that this crack occurred before Area III as no striations were detected in Area III. The fatigue crack growth at the crack tip in Area II suggest that this crack existed for some time as about 90,000 stress cycles would be required to propagate the crack 0.075 in. (1.9 mm). Crack Area II was completely contained inside the web and was not visible from the web surfaces. Hence this pop-in crack probably existed at the time of major fracture.

Crack Area III appears to have occurred during the restoration of the cracked girder. No fatigue crack growth is evident at the tip which would correspond to the growth observed in Area II. Fortunately the crack penetrated the inside web surface and was detected.

There are no obvious reasons why the second crack did not result in the girder fracture. For some reason the fatigue and pop-in crack were not as severe as the crack that developed at the fractured cross section. After the fracture developed, the girder would not be as highly stressed at the next floor beam and this likely permitted the subsequent fatigue crack growth that was detected in Crack Area II. Since the girder was repaired during warm weather, the web steel could resist the larger crack size without fracturing the entire girder when the cross section was restored and load reapplied.

SUMMARY

The fracture of a main girder of the Lafayette Street Bridge was due to the formation of a fatigue crack in the lateral bracing gusset to transverse stiffener weld. This fatigue crack originated from a large lack of fusion region which existed near the backup bars. At the failure gusset, this groove weld intersected the transverse stiffener-web weld and hence provided a direct path for the fatigue crack to follow into the girder web.

After the fatigue crack had nearly propagated through the girder web, it precipitated a brittle fracture of part of the web and all of the tension flange. Complete fracture was arrested by the tie plate which bridged the crack. Subsequent cyclic load permitted further fatigue crack growth over a limited time interval which eventually resulted in substantial extension of the web crack.

A second smaller crack removed from the girder showed comparable fatigue crack growth. The web plate was observed to just satisfy the fracture toughness requirements of the 1974 AASHTO Specification for Temperature Zone 2. The flange and gusset plates met the requirements for Zone 1. If other fatigue cracks are removed or properly arrested, the material in the Lafayette Street Bridge will provide adequate resistance to brittle fracture.

Since other lack of fusion discontinuities are probable in other gusset-transverse stiffener welds in the two bridges, all gussets located in regions of cyclic stress range and tensile stress should be treated to prevent and detect any fatigue crack growth into the girder web.

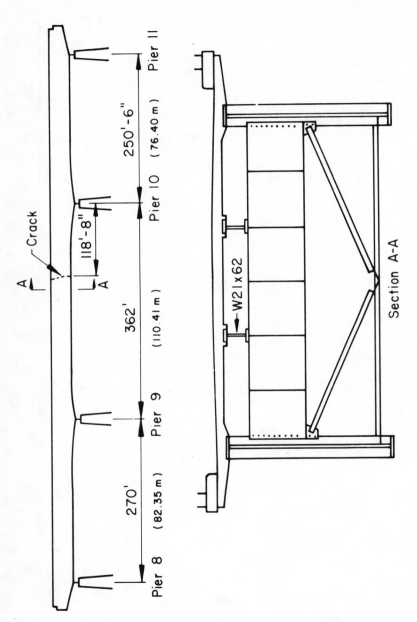

Fig. 1 Schematic of Span and Cross Section

Fig. 2 Cracked Girder of
        Lafayette Street Bridge

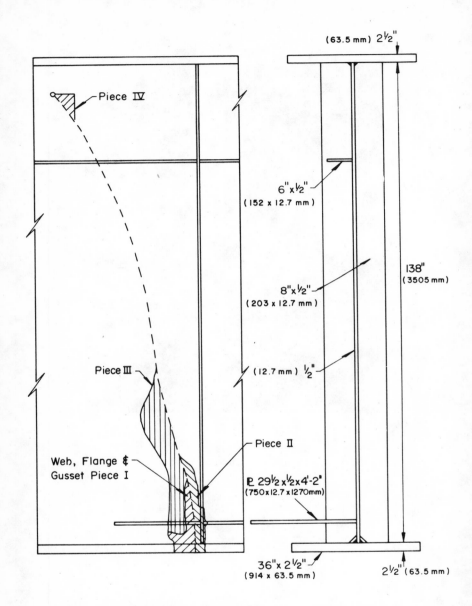

Fig. 3   Schematic of Girder Showing Crack and
Sections Removed for Examination

Fig. 4a Fracture Surface of Gusset to Stiffener Weld
and Adjacent Web - Piece II

Fig. 4b Fracture Surface of Gusset-Web Showing
Lack of Fusion - Piece II

Fig. 5 Fracture Surface of Web-Flange Boundary

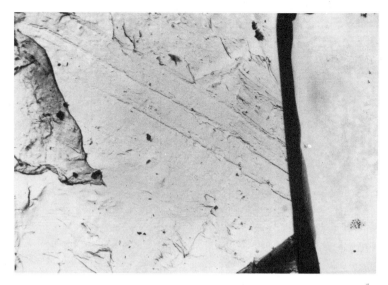

Fig. 6a   Crack Growth Striations in Region 1

Fig. 6b   Crack Growth Striations in Web

Fig. 7a Plug Removed from Web with Gusset Attached Cut to Expose Crack Surface

Fig. 7b Exposed Crack Surfaces in Plugs

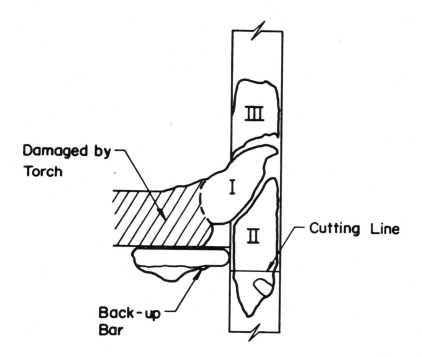

Fig. 8   Schematic of Crack Surface

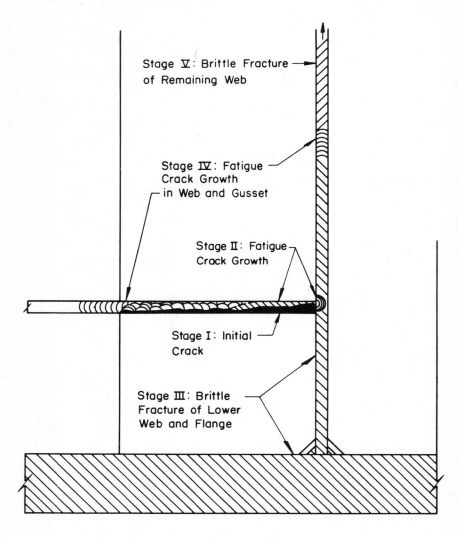

Fig. 9 Schematic of Stages of Crack Growth in
Gusset, Web and Flange

APPENDIX I

Corrective Action for Lafayette Street Bridge

The scheme shown in Fig. A1 was used to retrofit the Lafayette Street Bridge. Vertical holes were cut into the gusset plate as shown schematically. After the gusset material was removed, the web surface and gusset could be carefully inspected for crack penetration using dye penetrant on the exposed web surface and the edge of the gusset. If no crack indication was obtained, the detail can be regarded as adequate. When any indication of a crack was provided by dye penetrant on the exposed web or hole surface, additional holes were cut into the girder web with a 1¼ in. hole saw on the two diagonal lines from the opposite side of the web surface. These two holes will intersect and remove an elongated section of the web. After these plugs are removed, the web surface in these holes should be ground smooth and then the exposed web thickness again checked with dye penetrant to insure that the web crack does not extend below the area that has been removed. Should anything be detected, an additional portion of the web should be removed. After all material is removed the surfaces should be carefully ground smooth before the area is painted.

Figure A2 shows a suggested detail configuration for this type of connection should it be used in the future. Both gusset plate welds should be prevented from touching or coming too close to the transverse stiffener welds. This will prevent a path from forming into the web in case a crack develops in the transverse stiffener gusset weld. In addition, the gap reduces the restraint in the region.

Furthermore, it would be preferable to not use a groove weld with backup bar that is perpendicular to the bending stress as the gusset transverse stiffener weld is. These welds are likely to have lack of penetration and above average initial flaw sizes. As a result they offer very low fatigue resistance and are likely to crack.

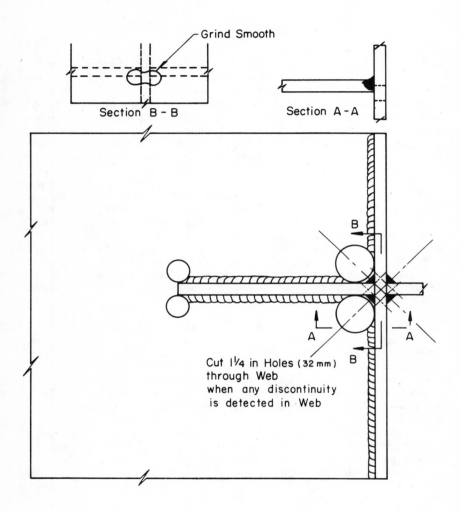

Fig. A1 Preferred Procedure for Corrective Action

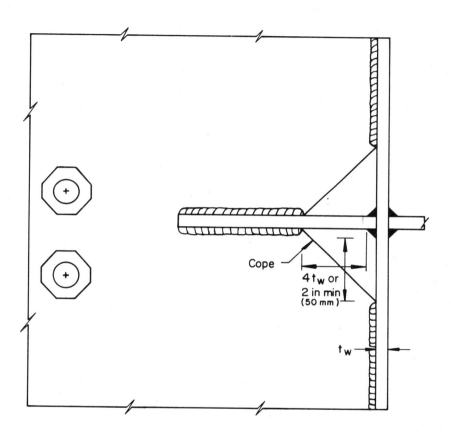

Fig. A2   Recommended Detail for Future Use

REFERENCES

(1) J. M. Barsom and S. T. Rolfe, "Correlations Between K_{IC} and Charpy V-Notch Test Results in the Transition-Temperature Range," ASTM STP 466, 1970.

(2) J. M. Barsom, "Investigation of Toughness Criteria for Bridge Steels," Applied Research Laboratory Report 97.018-001(5), U. S. Steel, February 8, 1973.

(3) ASTM, "Standard Method of Test for Plane-Strain Fracture Toughness of Metallic Materials, ASTM E399-74, 1974 Book of Standards.

(4) American Society for Metals, "Fractography and Atlas of Fractographs," Metals Handbook, 8th Edition, Vol. 9, ASM, 1974.

(5) J. L. McCall, "Fracture Analyses by Scanning Electron Fractography," MCIC 72-12, Metals and Ceramics Information Center, BaHelle-Columbus, Columbus, Ohio, 1972.

(6) J. W. Fisher, et al., "Fatigue Strength of Steel Beams with Welded Stiffeners and Attachments," NCHRP Report 147, Transportation Research Board, 1974.

(7) J. W. Fisher, "Guide to 1974 AASHTO Fatigue Specifications," AISC, 1974.

ACKNOWLEDGMENTS

This investigation and analysis was undertaken by the authors for the Office of Bridges and Structures, Minnesota Department of Highways.

The authors would like to acknowledge the assistance provided by Mr. K. U. Benthin, Bridge Engineer and his staff during this study. Their analysis of the structure provided the information needed to evaluate crack growth and their observations of the fracture were invaluable in tying the many pieces of the puzzle together.

A FAILURE ANALYSIS OF RAILROAD RAIL

E. I. Savage, R. K. Steele,
and J. M. Morris
U.S. Dept. of Transportation, Transportation Systems Center

There are over 650 reported train accidents each year in the United States attributed to broken rails, and the trend in recent years has been toward an increase in the number of occurrences. These accidents occur despite the removal annually of approximately 200,000 defective rails which are found by visual or nondestructive inspection. On this basis, considerable incentive exists to both better understand the mechanisms of rail failure behavior and to improve the effectiveness of inspection in an effort to reduce the broken-rail caused accident rate.

The types of rail defects most frequently found are depicted in Table I with a ranking of the percentage cause of derailments and frequency of detection. The action specified by Federal Regulation until the rail is replaced is also listed. Vertical and horizontal split head failures involve a progressive longitudinal fracture in the head of a rail where either failure has occurred in a x–z plane or an x-y plane, (Fig. 1). These defects, specifically, the vertical split head (VSH), may grow internally to be several feet in length before breaking through to the surface. Due to the almost totally embedded character of such a defect, its existence can be determined only by nondestructive inspection techniques prior to failure. Such a failure does not produce a fail-safe indication, as would a transverse rupture, in signalled territory. If such a failure occurs on the gauge side of the rail head (Fig. 1), train derailment is

29

Defect	Location/Shape	% Derail△/ % Detect	Type	Size	Action Required until Defective Rail is Replaced — Action
Transverse Type: · Transverse Fissure (TF)	EB / DF/CF	TF : 23%/10%	TF CF	HEAD AREA (a) <100%	10 mph MAX.
				(b) 100% (+)	VISUAL SUPERVISION
· Compound Fracture (CF) · Detail Fracture (DF) · Engine Burn Fracture (EB)		DF/CF: 8%/5% EB: 6%/5%	DF EB	HEAD AREA (a) <20%	30 mph OR LESS (UNTIL JOINT BARS ARE APPLIED AND THEN 50 mph OR LESS*
				20% to 100%	10 mph OR LESS* UNTIL JOINT BARS ARE APPLIED AND THEN 50 mph OR LESS*
				<100%	10 mph MAX.
				(b) 100% (+)	VISUAL SUPERVISION; OR JOINT BARS THEN 50 mph OR LESS*
Longitudinal Type: Vertical/ Horizontal Split Head	VSH / HSH	HSH/VSH: 26%/16%	VSH HSH	LENGTH (a) <2"	50 mph OR LESS;* INSPECT IN 90 DAYS
				(b) 2" to 4"	30 mph OR LESS;* INSPECT IN 30 DAYS
				(c) >4"	10 mph MAX.
				(d) CHUNK MISSING	VISUAL SUPERVISION
Rail End Type: Bolt Hole/ Head Web Separation	HW / BH	BH/WH: 20%/58%	BH HW	LENGTH (a) <1/2"	50 mph OR LESS;* INSPECT IN 90 DAYS
				(b) 1/2" to 3"(HW) 1-1/2" (BH)	30 mph OR LESS;* INSPECT IN 30 DAYS
				(c) >{3"(HW) 1-1/2"(BH)	10 mph MAX.
				(d) CHUNK MISSING	VISUAL SUPERVISION

△ Sperry Rail Service 1970-1974 FRA Accident Data   *Set by Class of Track

TABLE 1.   CATEGORIES OF MAJOR RAIL DEFECTS

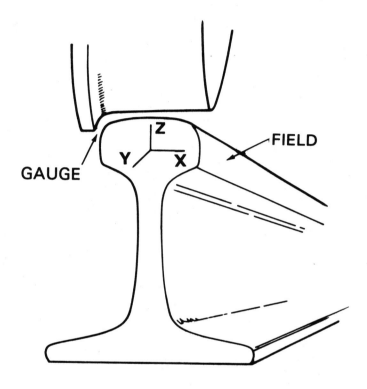

Figure 1. Rail/Wheel Geometry and Rail Head Coordinate Axes

virtually unavoidable. As will be discussed later, it is only under optimum conditions of defect position and ultrasonic probe arrangement that such defects will be detected when still small in size. The failure analysis which follows attempts to determine those factors responsible and thereby contribute to the development of safer rail usage. Improvements in the effectiveness of inspection practices to guard against such catastrophic failures are also described.

BACKGROUND

The specific vertical split head failure to be considered and its consequences are illustrated in Fig. 2. Here Amtrak Train #40 derailed on a 2° curve after the trailing truck of the sixth passenger car dropped between the rails. The speed limit on this section of track was 55mph. The failed rail was a 152 lb/yd section rolled in 1945. Its chemistry was within conventional specifications (1): 0.78 w/o C, 0.95 w/o Mn, 0.115 w/o Si, 0.0052 w/o O and 0.032 w/o S. The low O:S ratio of approximately 0.2 does not classify it as a steel susceptible to the damaging effects of MnO inclusions. The characteristic pearlitic microstructure of the rail is illustrated in Fig. 3.

The rail was sectioned for fracture toughness and microhardness tests. Average K_q values of 43 ksi $\sqrt{in}$ were obtained using both compact tension and precracked Charpy specimens. (K_q values are quoted since the compact tension specimen thickness was less, and the notch depth greater, than that specified in ASTM E399 for a valid K_{Ic} test.) This value of approximately 43 ksi $\sqrt{in}$ is approximately 12% higher than that reported for AREA continuously cooled rail (2) and may be associated with the use of K_q rather than true K_{Ic} values.

The microhardness readings varied from a Rockwell C scale reading of 55 near the surface (Fig. 4) to a value of 36 observed 0.5 in (12mm) below the running surface. Away from the hardened head of the rail the average hardness was R_c 25.

Figure 2. Derailment of Amtrak Train #40 Caused by Vertical Split Head Defect

Figure 3. Pearlitic Microstructure of Failed Rail. Etched with Nital, x500

Almost three years before the accident occurred the rail was transposed from the high side to the low side of the curved track due to excessive gauge corner wear. The pattern of wear and metal flow in the rail head resulting from service is shown in Fig. 4, in comparison to the profile of an unworn equivalent 155 lb/yd section. Approximately 9mm has been removed from the former gauge side of the head and 5mm from the current gauge side. In addition, the running surface had been severely reshaped which would lead to considerable eccentric loading when the rail was transposed due to a significant shift of the wheel/rail contact area to the gauge corner of the rail.

It is appropriate to comment briefly here on the service environment of rail, specifically the types of loading and stresses experienced. A more complete discussion is contained within several U.S. Department of Transportation reports (3,4,5,6). Fig. 5a shows that the stresses induced in a rail due to a passing vehicle are the combined results of three basic wheel/rail loading actions. When the vehicle approaches to within 6 to 12 feet of a particular point in the rail head, that location experiences a tensile bending stress due to the flexural action of the rail on the elastic foundation of the ties, ballast and subgrade. As the vehicle approach is closer to the point, the flexural tensile stress becomes a compressive stress roughly five times greater than the previous tensile stress. Significant variations in these flexural stresses occur due to the wide variations in load carrying capacities of individual ties. Nonuniform rail support in conjunction with large variations in the lateral position of the wheel on the rail head can lead to local bending and twist of the rail inducing subsequent variations in the flexural stresses in the head-web fillet regions.

Finally, when the application of the wheel load comes within 0.3 to 0.5 in (8-13mm) of a particular point in the rail head, large contact stresses develop due to the local deformation of the rail head near the region of application of the wheel load, as shown in Fig. 5b. In general

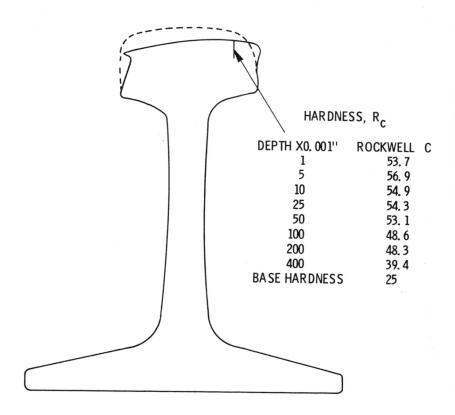

HARDNESS, R_c

DEPTH X0. 001"	ROCKWELL C
1	53. 7
5	56. 9
10	54. 9
25	54. 3
50	53. 1
100	48. 6
200	48. 3
400	39. 4
BASE HARDNESS	25

Figure 4. Cross Section of the Failed Rail (Solid Line) and Profile of a New 155 lb/yd Rail (Dashed Line)

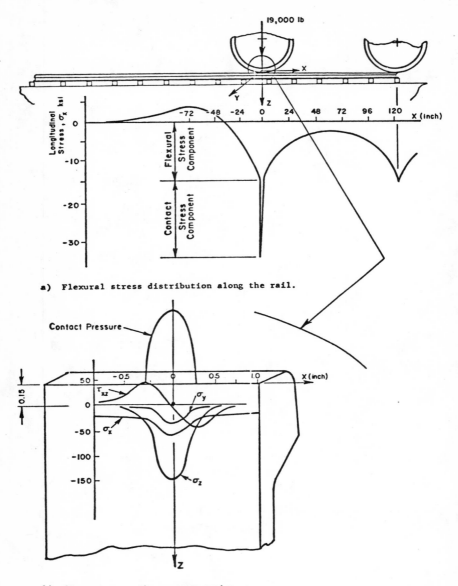

a) Flexural stress distribution along the rail.

b) Stresses near the contact region.

Figure 5. Stress History in Rail Head Due to Rolling Load (Ref. 3)

these stresses are much larger in absolute magnitude than are the flexural stress components and are usually compressive with the exception of a transverse shearing stress component which completely reverses as the rolling load passes. When wheel loads exceed approximately 700 lb per inch of wheel diameter, the resulting stresses in the rail frequently exceed the yield strength (7). This yielding of the rail head, to a distance of 0.15 to 0.30 in (4-8mm) below the running surface, produces a region of residually stressed material beneath the cold worked surface layer (Fig. 11). It is within the tensile region that many horizontal cracks and split head defects are believed to initiate (6).

Figure 6 shows the suspected origin of the VSH along with a narrow, 0.25 in (6mm) tall, horizontal crack which runs beneath the cold worked surface layer. Both the horizontal crack and the origin of the VSH lie within the region of residual stress development. Since the rail exhibited substantial metal flow both longitudinally and laterally at the running surface, it was anticipated that high residual stresses had developed internally. Since such stresses are difficult to measure accurately, it was decided to obtain qualitative information concerning the stress distribution through the use of a modified Sachs technique. For this purpose a rectangular solid specimen was machined from the rail. The maximum tensile stresses, both longitudinal and transverse, occurred at a depth of approximately 0.72 in (18mm), the depth at which the vertical crack is believed to have initiated.

From Fig. 7a it is observed that the horizontal crack extended almost the entire length of the VSH which was 4 ft (3.2m). The effect of this horizontal crack, although benign in itself, was most unfortunate as regards the circumstances of the derailment. It should be noted that the rail had been inspected by magnetic induction and ultrasonic techniques approximately two months prior to the accident. At this time numerous surface defects were noted and an indication of the horizontal defect was recorded. However, the latter defect was judged noncondemnable because of its narrowness. Unfortunately the origin of the vertical split head

Figure 6. Suspected Origin of the VSH (Left) and Horizontal Crack
Extending to Right

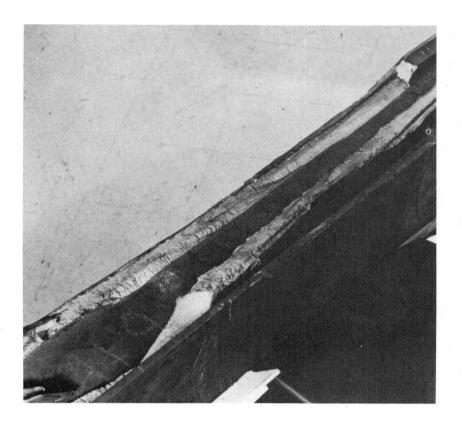

Figure 7a. Fracture Surface of Vertical Split Head

was not detected. It appears that the badly spalled running surface of the rail could have sufficiently attenuated the defect indications to prevent resolution of the VSH origin. Also the off-center alignment of the VSH would have made it difficult for conventional ultrasonic probe arrangements to detect without the incorporation of a side-scanning probe or the use of probes coupled in a pitch-catch mode.

FAILURE ANALYSIS

Based on observations of the character of the fracture surface and knowledge of fatigue crack growth behavior in rail steels (5b), the problems of defect initiation and driving force for the growth of vertical split heads will be considered. The origin appears connected to the horizontal crack which runs the length of the VSH at a depth of 11-18mm in a region of high tensile stress. The horizontal crack surfaces were covered by a shiny black oxide, probably a fretting corrosion product (Fig. 6). The vertical crack exhibited no black coating. Examination of the fracture surface (Fig. 7b) illustrates the presence of several distinct rust "rings" which appear to represent contours of the crack tip during progressive stages of defect growth, indicating that the crack was effectively arrested or stopped long enough for an iron oxide deposit to develop. It should be noted that the moisture necessary for rust formation is conjectured to have been supplied from the outside environment through the horizontal crack to the vertical split head. Capillary action effectively drew the moisture into the vertical crack tip area.

Scanning electron microscopy (SEM) was accomplished at the origin and at several locations along the progression of the vertical crack to investigate metallurgical causes for the failure. The origin region exhibited a striated "woody" structure as shown in Fig. 8. Such a structure has been previously reported to be correlated with Mode II or shear crack growth (8). No localized region of high inclusion content could be found and

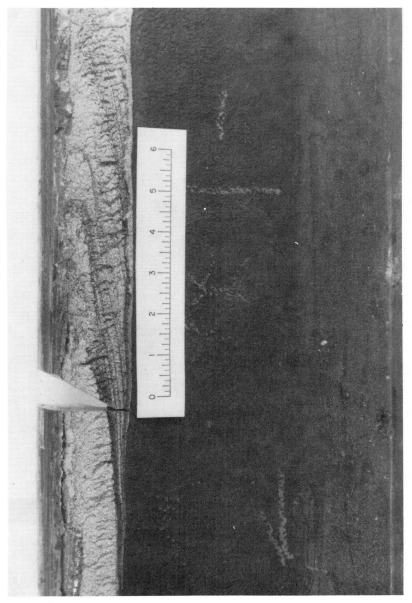

Figure 7b. Rust Rings on VSH Surface. Note Spalling on Running Surface at Top. (Samples Removed from Middle)

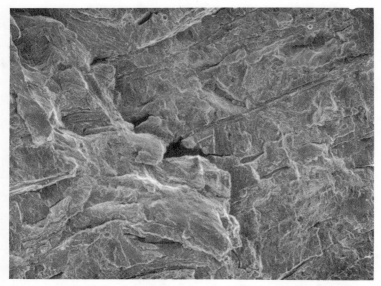

Figure 8. SEM View of Woody Structure Observed at Origin of VSH
x100

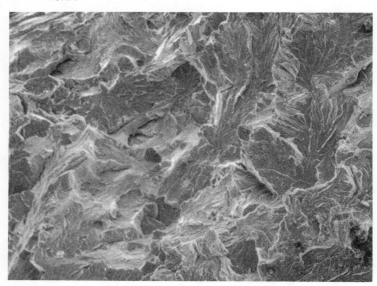

Figure 9. SEM View of Cleavage Facetting Observed Away from
Origin. x100

identified as the likely cause for initiation of
the vertical crack. In regions away from the origin, SEM examination reveals a uniformly high percentage (95%-98%) of cleavage facetting throughout
the length of the vertical crack (**Fig. 9**). Based
upon the results of previous research (5c, 9, 10),
such a large percent of facetting is associated with
very high stress intensity factors, approaching K_{Ic}
or K_{Id} and hence high crack growth rates (10^{-3} in
per cycle). Such growth is generally considered
to proceed in the Mode I or crack opening mode, and
the applicability of experimentally derived K_I values to predict growth rates for the VSH defect is
discussed below.

Based on experimental investigations (3,6),
the flexural component of K_I on the bottom of a
0.60 in wide VSH defect, subject to a central vertical load of 19kips, is found to be approximately
4 ksi$\sqrt{in}$ when the defect is centered 0.75 in below
the running surface. The mixed mode contributions
for this defect are a K_{II} component of approximately
1 ksi$\sqrt{in}$ and a K_{III} component of zero. In determining the effect of mixed modes on the growth of
VSH defects we will adopt an analysis developed by
Besuner (11) for transverse defects in rail. This
approach utilizes an effective stress intensity
factor

$$\Delta K_{I\ eff} = (\Delta K_I^2 + c\Delta K_{II\ eff}^2)^{\frac{1}{2}} \qquad (1)$$

where

$$\Delta K_{II\ eff} = (\Delta K_{II}^2 + 1.43\Delta K_{III}^2)^{\frac{1}{2}} \qquad (2)$$

The value of c is predicted to be between 1 and 3.
Thus, for values of K_I of 4 ksi$\sqrt{in}$, K_{II} of 1 ksi$\sqrt{in}$
and K_{III} of zero, one obtains a value for $\Delta K_{I\ eff}$
of approximately 4.2 ksi$\sqrt{in}$ which is too low to
produce significant crack growth and is below the
experimentally observed threshold for rail steels
of 10 ksi$\sqrt{in}$ (5). Since the growth rates of the
VSH inferred by fractographic examination suggest
values that are several orders of magnitude greater
than threshold growth levels, sufficiently high K_I

levels must exist to cause the defect to grow.
To rationalize this discrepancy between the observed
$(10^{-3} - 10^{-2}$ in/cycle) and the calculated growth
rates the contributions of transverse residual
stresses, the effects of which may be magnified by
eccentric loading, must be considered.

For the most severe case studied (6), relating
to the 0.60 in wide VSH, the presence of a combined
residual stress field and a 19 kip applied vertical
load results in a K_I of approximately 12 ksi$\sqrt{in}$, a
value still too low to produce appreciable crack
growth. However, such an increase in stress intensity factor value, due to the presence of residual
stresses, does illustrate their influence on crack
growth behavior.

Accordingly, we will examine the potential
effects of residual stress levels to account for
the observed high crack growth rates in the VSH
under study. Broek and Rice (5b) have observed
that rail steels fit a growth law in terms of
mean stress

$$da/dN = C(1-R)^2 \left(\frac{K_{max}^2 - K_{th}^2}{K_c - K_{max}} \right) \cdot K_{max}^n \quad (3)$$

where da/dN is the crack growth rate, C, n, and K_{th}
(threshold stress intensity factor) are constants
which do not depend on R. The stress ratio, R, is
influenced by the mean stress which, in turn, is a
function of the residual stresses. Thus, assuming
a fixed ΔK of 16 ksi$\sqrt{in}$, if K_{th} were 10 ksi$\sqrt{in}$,
n were 3 and K_c had a value of 50 ksi$\sqrt{in}$, a 3:1
increase in mean stress (residual stresses corresponding to a ΔK increase of 24:8 ksi$\sqrt{in}$) would pro-
duse a twenty-two fold increase in crack growth
rate as illustrated in Fig. 10.

Despite the crudeness of our calculations
the effect of residual stresses has been demonstrated with regard to increasing fatigue crack
growth rates and thus, decreasing safe rail life.
The presence of higher mean stresses in rail due
to normal traffic where loading may occasionally

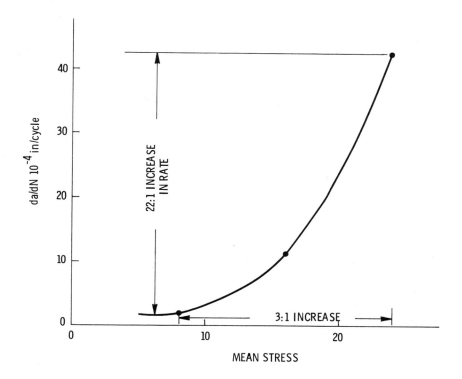

Figure 10. Effect of Mean Stress on Crack Growth Rates for Rail
Steels. (Based on Ref. 5b)

reach levels of 40 or 60 kips will result in increased crack growth. For the specific rail of interest, the eccentricities in loading caused by transposition of the rail, together with the excessive wear it exhibited, can be assumed to be intimately linked with high mean stress and residual stress levels. Due to these factors crack growth levels would approach those observed and residual stresses may be assigned as the controlling factor in the defect growth rate since it is a function of mean stress levels, which, in turn, depend on residual stresses.

CONCLUSIONS

On the basis of this investigation it is felt that the high mean stress levels resulting from the combined effects of an excessively worn rail which had been transposed in an effort to prolong its useful life, together with eccentric loading, were the major causes of rail failure, since fractography of the rail revealed no metallurgical cause for the initiation of this failure. It is noted that the vertical crack initiated at a level beneath the running surface where surface strain change measurements indicated the presence of high residual tensile stresses. From our calculations and the work of other researchers, highly probable loads (in the vicinity of 19 kips), in the absence of residual stresses, are insufficient to cause appreciable crack growth. However, the joint presence of both applied wheel/rail loads and residual stress fields are likely to be responsible for crack growth.

In addition to the vertical defect a long but narrow horizontal crack developed under the cold worked region. A dark oxide on its surface suggested that it existed long before the vertical crack had initiated, and its presence may well have obscured the origin of the vertical crack from detection. On this basis it should be understood that rails which exhibit severe wear, especially those which have been transposed, are considered more likely to develop defects due to eccentricities in loading and residual stresses and should be inspected according to more stringent requirements than those specified for less worn rail. Since the running surface of worn rails may pose difficulties in obtaining accurate detection signals, it is recommended that a side scanning probe should be used on worn rails. Also indications of "non-condemnable" defects should be further investigated in such worn or transposed rails to assure a safe rail usage.

Finally, through experience it is recognized that little correlation has been found between the conditions necessary for the occurrence of vertical split heads and rail service environment parameters. Such defects occur on both tangent and curved track, on new and old rail, and on passenger and freight lines. Recent work sponsored by the United States Department of Transportation at the Army Mechanics and Materials Research Center has suggested that the stress state necessary for Mode I growth of a vertical split head, from an initially small crack, may be intimately related to the tie spacing. The applicability of such a suggestion to rail in service appears relevant because examination of the bottom of the failed rail discussed in this analysis revealed that the VSH origin occurred just ahead of a crosstie.

REFERENCES

(1)   AREA - Manual for Railway Engineering - Specifications for Rail Steels - 1975.

(2)   AREA - Engineering Aspects of Current Rail Sections, Report No. ER-15, Sept. 1961.

(3)   Johns, T. G. and Davies, K. B.; Preliminary Descriptions of Stresses in Rail Steels - Interim Report; FRA-ORD-76-294, Nov. 1976.

(4)   Prause, R. H., et al; Evaluation of Analytical and Experimental Methodologies for the Characterization of Wheel/Rail Loads, DOT-TSC-1051 Interim Report, April 1976.

(5a) Broek, D. and Rice, R. C.; Fatigue Crack Growth Properties of Rail Steels, DOT-TSC-1076 Final Report (Part I), July 1977.

(5b) Ibid; Prediction of Fatigue Crack Growth in Rail Steels, DOT-TSC-1076 Final Report (Part II), Sept. 1977.

(5c) Buchheit, R. D. and Broek, D.; Microstructural and Fractographic Characterizations of Fatigue Tested Rail Steels, DOT-TSC-1076 Final Report (Part III), Oct. 1977.

(6) Johns, T. G., et al; Engineering Analysis of Stresses in Railroad Rail: Phase I, DOT-TSC-1038 Final Report, June 1977.

(7) Stone, D. and Steele, R. K., Proceedings ASTM Symposium on Rail Steels, Denver, Colo., November 1976.

(8) McEvily, A. J. - Private communication to authors.

(9) Fowler, G. J.; Fatigue Crack Initiation and Propagation in Pearlitic Rail Steels - Ph.D. Dissertation - UCLA, 1976.

(10) Beevers, C. J.; Improving the Fatigue and Fracture Resistance of Steels. Report prepared for Transportation Systems Center, U.S. D.O.T., June 1977.

(11) Besuner, P. M.; Fracture Mechanics Analysis of Rails with Shell-Initiated Transverse Defects, Failure Analysis Associates, Palo Alto, Nov. 1975.

THE ROLE OF NDT
IN AIRCRAFT ACCIDENT INVESTIGATIONS

R. B. Davis
Federal Aviation Administration
Atlanta, Georgia

Some form of non-destructive testing is used during the investigation of practically every aircraft accident or incident to assist the parties involved in determining the probable cause. Some of these methods involve very exotic and expensive equipment, others are comparatively simple and are based on the background knowledge and experience of the investigators involved.

I plan to review a number of aircraft accidents and incidents and explain some of the testing done during the investigation.

Several years ago, a twin-engined airplane, flown by a well qualified pilot with a nurse as a passenger, departed a controlled airport in Mississippi in instrument conditions. The pilot was asked to report leaving 7000 feet, which he did. This was the last transmission from the aircraft. The wreckage of the airplane was located three days later.

Shortly before I arrived, an FAA Designated Aviation Medical Examiner completed a preliminary autopsy of the pilot. Concerned with his findings, he packaged several vital organs and stored them in liquid nitrogen.

After arranging with an airline to deliver these parts to Washington, they were re-packed in dry ice and rushed to the airport. The package was picked up in Washington and hand carried to the Armed Forces Institute of Pathology where their examination confirmed the results of the preliminary autopsy.

The findings of both the aviation medical examiner and the
Armed Forces Institute of Pathology disclosed that the pilot
died from carbon monoxide inhalation. Further, even though
the pilot's compartment burned after the crash, there was no
evidence that the pilot at any time breathed in any flames.

The combustion heater was then removed from the wreckage
and shipped to one of the leading metallurgical laboratories
in this country. A detailed examination disclosed three minute cracks in the combustion can which would have, in their
opinion, emitted enough carbon monoxide to kill the pilot.

Some years ago, a helicopter engaged in crop spraying
operations in North Carolina, while attempting a takeoff from
a paved road, encountered ground resonance phenomena and lit-
erally shook itself to pieces. What, at first.glance, made
this accident very unusual was the fact that the pilot had
made at least a dozen takeoffs that morning with no sign of
a problem.

The pertinent components were removed and carried to a
nearby Army maintenance facility for tear-down and examination. Three things were learned:

1. The operator had installed a mis-matched set of rotor
    blade dampners.

2. The dampners installed were not torqued to the limits
    established by the manufacturer, and,

3. The skid shock strut pressures were outside the specified
    tolerance.

Why was the pilot able to operate all morning with no
indication of trouble? In talking to the pilot, we learned
that on all previous landings to refill the spray tanks, he
landed in the cotton field itself. He had been operating
"on the ragged edge" so to speak, and the soft ground had
absorbed enough vibration to keep him out of trouble. When
he attempted a takeoff from a hard surface, this very thin
margin of safety no longer existed.

In the mid-60's, an incident occurred northwest of Baltimore involving an executive jet that resulted in one of the
more extensive investigations to be conducted by the Federal
Government and the manufacturer of the airplane involved.

The aircraft, while cruising at 15,000 feet, suddenly pitched down into an uncontrollable dive. Control was finally recovered at an elevation of 2000 feet between two hills of approximately 2600 feet elevation. Upon landing at an airport near Baltimore, it was discovered that a number of fairings and wheel well doors were missing, plus about six feet of the right hand stabilizer and the entire right hand elevator!

During the course of this investigation, which lasted about two years at a cost of approximately one million dollars, every existing NDT method was employed at one time or another. The pitch control system was re-evaluated in great detail with emphasis on the hydraulic boost assembly. The flutter characteristics were re-evaluated through a series of stiffness tests and wind tunnel model tests. Analytical studies and wind tunnel tests were conducted wherein control malfunctions were simulated and resultant flight characteristics obtained. Detailed metallurgical analyses were made on every failed part. Aerodynamic loads were re-computed and the horizontal stabilizer re-static tested.

Two very interesting observations were made during this rather lengthy investigation. While conducting Rockwell hardness tests on the aft horizontal stabilizer box beam caps, it was discovered that the tensile strength of two of these members, at the root end, was approximately 50% of that required by the drawings. I happened to be in the laboratory when this deficiency was noted and immediately took steps to suspend the airworthiness certificates on all aircraft of this make and model. Telegrams went out to all owners that evening at 7:00 p.m. to ground their airplanes.

Now, why were these beam caps understrength? The heat treat furnace used in the heat treating process was approximately two feet shorter than the beam caps, so the root ends were left sticking outside the furnace. Additional tests were run, however, with a twelve inch section of cap removed and the stabilizer still met all ultimate strength requirements.

The second interesting discovery was made by a state crime lab. Splatters of a dark reddish substance were found on the stabilizer front beam web at the break area. Since crime labs are supposed to be experts at detecting blood, they were asked to analyze these splatters. Three days after delivering the parts to them, I received a call from the assistant director. He reported that the substance on the front beam web gave a positive reaction to their reagent and it was definitely blood. He emphasized that blood was the only known substance that would cause a positive reaction to this particular reagent.

Several days later, he called back and reported one more substance that would give a positive reaction--zinc chromate. I asked him how many airplane painters he had sent to the electric chair, and he didn't think that was funny at all!

An example of non-destructive testing on a very small scale took place in one of our general aviation district offices some years ago. This was not done as a result of an airplane accident, but it may have prevented one.

A gentlemen came into the office and told one of our inspectors that every time he flew his airplane, he experienced a severe headache and nausea. He asked if the FAA would check his airplane for presence of carbon monoxide. This was done, and the CO content of the air in the cabin in the most critical configuration was found to be negligible.

A few weeks later, the man was back again complaining of symptoms of carbon monoxide poisoning when he flew his airplane. His plane was checked again with negative results.

After the man came to the office for the third time, the inspector pinned a small carbon monoxide detector button on the man's shirt. The man said he was on his way to the airport where he kept his airplane, some 20 miles away. The inspector made arrangements for the airport manager to meet him and remove the detector button.

When the man reached the airport, the detector had turned black, an indication that it had been in contact with a very high concentration of carbon monoxide. Seems the man was driving an old car with a faulty muffler and was feeling the effects of the CO only after he climbed to a less dense atmosphere.

Another example of testing on a small scale involved a single engine, four place airplane which departed an airport in central Florida in the late 60's. At the controls was a 200 hour private pilot. With him were his wife and small child.

Immediately after takeoff, the aircraft entered a shallow left bank. Power was reduced and the aircraft crash landed less than 100 feet from the terminal building, narrowly missing a loaded commercial airliner. The airplane hit a shallow bank, ripped off the landing gear and skidded 65 feet before coming to a stop. Fortunately, there were no injuries.

When our inspector arrived on the scene, the pilot told him that when the aircraft started to turn left, he attempted to stop the turn by applying right aileron, but the controls were jammed. The inspector checked the aileron control system from the cockpit and found it to be operating normally. The pilot insisted it was jammed when he attempted his takeoff, and since I was in Miami at the time, was asked to stop by on my way home and shake down the aileron system.

Upon removal of the right wing tip, a bucking bar was discovered in the outer wing bay. It was learned that the bucking bar, when placed between the aileron bellcrank and a wing stringer, would restrict the control movement in exactly the same manner as that described by the pilot. As it turned out, rather than being blamed for the accident, the pilot involved was praised for the expert way in which he handled the situation.

Another example of the use of NDT in determining probable cause of an airplane accident involved a four engine turbo-prop transport. The aircraft, enroute from California to Delaware, lost radar and voice contact with the Kansas City Air Route Traffic Control Center shortly after a request was made to deviate around thunderstorm activity ahead on it route of flight. Witnesses later reported pieces of wreckage falling from the aircraft just before impact, and some witnesses noticed that either a wing or a wing and engine were missing.

Since the outer left wing was recovered almost two miles from the main impact site, this area of failure was examined in detail. Metallurgical examination of the fractured surfaces revealed that the lower front spar cap fractured completely in fatigue. The spar cap was deformed at the primary origin area of the fatigue fracture. Hardness and electrical conductivity of the spar cap material were normal for the type material involved.

The lower portion of the front spar web contained an approximate 4.9 inch fatigue crack with intermittent tensile tearing.

The lower wing skin fracture stemmed from pre-existing fatigue cracks at the first fastener hole located 3/4 inch outboard of the primary origin area at which the spar cap failed. Deformation and multiple cracks were noted at the origin of the skin fatigue fracture in the lower wing.

With the above information in hand, the FAA immediately issued an airworthiness directive requiring an initial and repetitive inspections of the affected area until it had been repaired or reinforced in a manner acceptable to the FAA.

Another accident involving an aircraft of the same model as that just discussed occurred north of Fairbanks, Alaska. The airplane was descending in moderate turbulence when, according to two witnesses, it broke up in flight. Since this in-flight break up happened just five months after the first wing failure, all fractured areas involving the wing were removed and immediately shipped to the National Transportation Safety Board metallurgical laboratory in Washington for analysis. The cockpit voice recorder and the flight data recorder were also recovered and sent to the NTSB laboratories for evaluation. A detailed metallurgical examination of all fracture surfaces failed to disclose any evidence of fatigue that could have contributed to the accident. A review of the information on the flight data recorder at the time of break up, and for several minutes preceeding, disclosed no information that was felt to have any real significance. The playback of the cockpit voice recorder revealed nothing except that the crew apparently had no warning of break up.

The NTSB has not yet released the probable cause on this accident; however, based on the latest evidence, it would appear to have been the result of a maintenace discrepancy.

One of the more interesting investigations I've had the opportunity to participate in took place in England last fall. The aircraft involved was a U.S. Air Force transport and I was invited by them to provide technical assistance.

While making an approach for landing in intermittent instrument conditions, the airplane was observed by a number of people to break up in flight. The major portions of the aircraft were confined to an area of approximately 1 1/4 miles in diameter.

At no time prior to the accident did the crew give any indication they were having problems. They were being vectored into position for landing by a remote radar facility and the crews last transmission to radar was that they thought they could remain VMC, or in visual contact, if they altered their course slightly.

Altitude at time of break up was less than 10,000 feet and speed was 250 knots, or less.

Since ligtning was a prime suspect, the entire surface of
the airplane was examined for evidence of a strike by two
lightning specialists from the Air Force, as well as two experts in that field from England. No evidence was found to
indicate a lightning strike.

All balance weights and structure adjacent to them were
examined for evidence of flutter. None was found.

The inboard portion of the right wing was reassembled on
a fixture and each and every failure on the spars, risers
and planks was examined with a magnifying glass. This metic-
ulous examination disclosed no evidence of any pre-existing
or fatigue failures.

Smears made by both metallic and non-metallic substances
were found on the upper surface of the fuselage and on the
lower portion of the vertical fin. These smears were cut
out and sent to laboratories in the U.S. for analysis.

There was no cabin or cargo compartment debris, such as
jackets, papers, cups, shoes, parachutes, etc., found along
the airplanes route of flight.

Very detailed and comprehensive autopsy reports were prepared for each of the crewmembers and passengers. The primary purpose of these examinations was to determine whether
fire was present prior to break up.

Parts were examined for evidence of inflight explosion.
None was found.

During my very first visit to the accident site, I discovered an Air Force fighter nose gear assembly under a
pile of rubble. My first thoughts were that we had a mid-
air collision on our hands and I could pack up and go home.
Then someone ruined my entire day by telling me the cargo
manifest included numerous aircraft parts!

In spite of that setback, the possibility of a mid-air
collision was investigated with negative results.

In summary, a large group of people spent two weeks or
more in and around a muddy sugar beet field in England
looking into every known probability that could result in
an inflight structural break up. Nothing was found that we
could, beyond a shadow of a doubt, point our finger to and
say, "that caused the loss of an airplane and many lives."

Since I was not an official member of the accident review board, I have never seen the report prepared by the Air Force. I do have some ideas, and if anyone wants to talk about it later, I'm available.

I hope this presentation has, in some small way, pointed out the very important role that testing plays in the investigation of aircraft accidents.

Now, if anyone has any questions, I'll be happy to try to answer them.

A PRACTICAL APPROACH TO FRACTURE
AND FATIGUE CONTROL OF CRITICAL PARTS

R. L. Circle
Lockheed-Georgia Company
Marietta, Georgia

ABSTRACT

Special controls on quality are required
to meet the objectives of the latest Air Force
design requirements. The need for such requirements are well founded in experiences of structural problems with many of the past generation
of military aircraft. The challenge is to implement special controls and upgrade quality
without impacting schedules and driving costs
up significantly. A program was developed by
the Lockheed-Georgia Company to supplement the
design and manufacture of a new wing box for
the C-5A. The C-5A Fracture and Fatigue Control
Plan outlines this program which emphasizes
selective identification of critical parts and
requirements, maximum use of existing techniques,
improved coordination between Engineering and
Quality Assurance, and Manufacturing, and added
closure to monitor the various system functions.

INTRODUCTION

Since 1969-1970 there have been significant changes in
the Air Force structural design philosophy. This new philosophy is strongly influenced by serious structural problems experienced in many older service aircraft as well as
some new system development programs. These problems have
resulted in high cost maintenance programs and in some
cases, serious safety problems. Many of these problems
are now well documented and References 1 and 2 contain
excellent reviews. The old design philosophy emphasized

57

undamaged structural static strength. Structure was assumed to be free of poor workmanship and a safe-life
(fatigue) methodology was used to evaluate life. Damage
tolerance was sometimes ignored; however, at best the
civil transport "fail-safe" approach[3] was used to provide safety with major structural damage present in the
airframe. Experience has shown many limitations to this
old design philosophy. Some of the more important of
these are:

    (a)   Materials with the highest static strength
          many times have relatively poor toughness
          when damaged.

    (b)   Even the most careful manufacturing and
          quality program cannot eliminate all
          sources of damage.

    (c)   True damage tolerance involves additional
          factors such as inspectability.

The following paragraphs address this new philosophy
and some of the resulting detail requirements. An example
of a comprehensive Fracture and Fatigue Control Plan is
outlined as an approach in the implementation of the
requirements.

GENERAL REQUIREMENTS

What does the new philosophy involve? The new philosophy is now well documented and some good examples can be
found in References 4 and 5. The major aspects of this
philosophy are:

    (a)   The design must assume that flaws will
          exist in new structure and the growth
          of these initial flaws must be evaluated.

    (b)   The size of these assumed initial flaws
          depends on inspection techniques, proof
          testing, or data from full scale tests.

    (c)   The design requirements are weighted by
          the inspectability of the design.

    (d)   An overall program must be defined such
          that quality is consistent with the design
          assumptions. This program is documented
          in a Fracture and Fatigue Control Plan.

The detailed manifestation of this philosophy and the precise requirements are contained in several new Air Force specifications. Three of the more significant of these are:

(a) MIL-STD-1530,[6] which outlines the overall airplane structural integrity requirements,

(b) MIL-A-83444,[7] which outlines the damage tolerance requirements,

and

(c) MIL-I-6870C,[8] the NDT requirements.

The development of these new ASIP (Aircraft Structural Integrity Program) specifications have resulted in significant new design requirements for the B-1, A-10, F-16, and the C-5A Wing Modification Program and are being imposed on all new programs. These new requirements include a full scale test program to verify that the design complies with the requirements prior to a decision on production go-ahead.

THE CHALLENGE

The greatest challenge is not as much to meet the durability and damage tolerance requirements, but to do so by implementing the necessary controls and quality without adversely impacting schedules and driving costs up significantly. The technical dimensions of the requirements are within reach of the current aerospace state-of-the-art. This includes durability and damage tolerance analysis, nondestructive inspection, availability of improved materials with controlled damage tolerance properties, and corrosion control. A Fracture and Fatigue Control Plan is required by Reference 6 and the approach this Plan defines will determine how well the challenge is met.

GENERAL APPROACH

The overall objective of a Fracture and Fatigue Control Plan is to define and integrate tasks that will:

(a) produce minimal maintenance critical structure - this aspect is closely identified with the prevention of general fatigue cracking that could result in maintenance problems and force down times,

(b)   prevent in-service failure of critical structure -
      this is closely associated with safety by con-
      trolled (or slow) flaw growth. This involves
      the application of damage tolerance design re-
      quirements; i.e., the fracture mechanics approach.

Figure 1 illustrates the overall concept of what is meant
by fracture and fatigue control. It is this overall coordination and integration of the disciplines and functions shown,
all of which are intimately involved at one time or another
with important aspects of critical structure. The Fracture
and Fatigue Control Plan is the primary integrating and coordinating force in this control. As is readily apparent
from the figure, virtually all organizations are involved.
Herein lies a major challenge - all organizations are involved and are interdependent in achieving the objectives
of the program. All organizations must work together in a
positive manner. Requirements must be realistic and manufacturing must step up to their responsibilities to upgrade
the product in critical areas to meet these requirements.
Most of the tasks involved in a plan are normal to past and
present development efforts. However, the new detail requirements impose the need for tighter controls and a coordinated effort of all the disciplines and organizations
involved. If the approach is not practical and achievable,
it simply will not work.

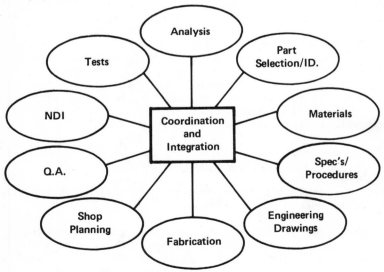

Fig. 1.   Concept of Fracture and Fatigue Control

At the onset of the development of the C-5A Wing Mod Fracture and Fatigue Control Plan, some guidelines were defined in order to outline the general approach to be followed. These general guidelines were:

(a) As much of the existing system (drawings, procedures, etc.) as possible would be used.

(b) The special controls would be limited to truly critical parts and features.

(c) Positive cooperation at all levels of all involved organizations was vital.

(d) Requirements must be realistic..

(e) Any changes or additions to the established system must be implemented prior to the time they were needed. This was an additional challenge since the go-ahead for the development of the plan was concurrent with the design go-ahead.

THE FRACTURE AND FATIGUE CONTROL PLAN

The overall C-5A Wing Mod Fracture and Fatigue Control Plan is divided into four (4) parts:

(a) Phasing

(b) Management

(c) Critical Part Selection Criteria

(d) Critical Part/Assembly Control System

The phasing is consistent with the four phases of the basic modification program.

PLAN MANAGEMENT CONCEPT

The basic concept of the plan management is to achieve closer coordinated and more intensified application of existing techniques, requirements, and procedures. The implementation and functions of the plan are accomplished and managed within each normal area of responsibility.

A Fracture and Fatigue Control Board is defined and is the primary management tool that provides overall coordination and monitoring of the total operation. This Board reports to the C-5A Wing Mod Program Manager. A Lockheed-Georgia Management Directive defines the overall plan in conjunction with the contract requirements and assures the overall support of the total facility.

CRITICAL PART SELECTION CRITERIA

Figure 2 depicts the criteria by which critical parts are selected. This process is basically analytical. The first question asked is:

"Is the structure safety of flight?"

(The Air Force definition of safety of flight is "that structure whose failure could cause direct loss of the aircraft, or whose failure if it remained undetected could result in loss of the aircraft".) The term "structure" as used here is considered to be both detail parts or major elements of structure, such as wing surface. If the answer is no, then the structure is non critical and the control plan is not applicable. If the answer is yes, then the structure is considered critical and falls into one of two categories.

If the structure will meet the design requirements with special material controls and normal quality, then it is classified as Category II. If special quality controls are required in addition to material controls to meet the design requirements, then the structure is classified as Category I. Initial flaw sizes are used as the discriminator between Category I and II. Structure can also be classified as Category I on the basis of prior experience with similar structure. Category I can be considered as that structure that is very closely designed by the fracture and fatigue requirements; i.e., the material or the stress level is set by the damage tolerance or the fatigue design requirements.

This criteria has worked out well and currently most of the wing lower surface major members and splice plates on the inner two-thirds (2/3) of the span fall into the Category I class. The upper surface and the outer one-third (1/3) of the lower surface are considered Category II.

Note that throughout this plan no discrimination is made between fracture or fatigue. The requirements and the controls necessary for both are considered so interdependent that it was felt to be unnecessary and in fact undesireable to try to discriminate.

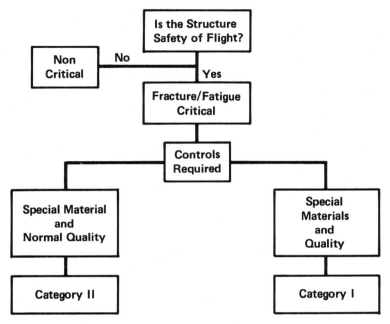

Fig. 2.   Fracture/Fatigue Critical Part Screening Criteria

CRITICAL PART CONTROL CONCEPT

The concept of what is meant by critical part control means special control of:

(a)   all operations and processes that could degrade material properties

and

(b)   processing that could have detrimental effect on meeting the design requirements - such as machining, assembly, handling, etc.

Both categories of critical parts are involved with the first item; only Category I is involved with the latter item.

Figure 3 is an illustration of the major significant features of the critical part control system. Requirements, particularly Engineering requirements, were completely reviewed and updated. In this case, updating also included reducing some requirements where they were found to be unnecessarily stringent.

Special identification procedures were developed to increase awareness of the special nature of the parts at all stages of the system - from the Engineering Drawing to the finished installation.

In-process controls were added in cases where existing controls needed supplementing.

Several key closure items were added to increase the basic system feedback - primarily between diverse operations such as Materials Engineering and Production.

An improved coordination system for everyone involved was implemented. The primary factor in improving coordination has been the Fracture and Fatigue Control Board where person-to-person contact has resulted in optimum approaches being defined.

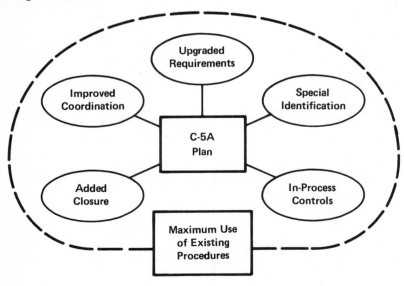

Fig. 3.   Significant Features of Control Plan

Last, but not least, in the putting together all the pieces of this system, the guideline of making maximum use of existing procedures has been followed.

IMPLEMENTATION ACTIONS

With a proper general approach and good practical guidelines established, the success of the plan depends on the detail implementation actions taken. The implementation actions will be discussed in the following paragraphs by five (5) categories shown in Figure 3.

Improved Coordination

Figure 4 illustrates the action taken in the area of improved coordination.

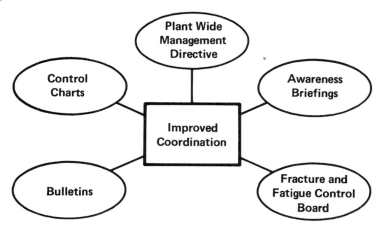

Fig. 4.   Implementation Action - Improved Coordination

A Lockheed-Georgia Company Directive was written that covers the plant wide implementation of the plan. This directive is in effect and assures the overall support of the total Lockheed facility. Awareness briefings on this plan have been given to personnel involved with the plan to help make them aware of the importance of the plan and the part they all play in the successful implementation. These have been given to approximately 200 people at all levels of management in Engineering, Quality Assurance, Manufacturing, and the major Subcontractor. A Fracture and Fatigue Control Board has been established and has

been meeting since the overall plan was approved by the
Air Force. A system of Fracture and Fatigue Control
Bulletins was originated. These are special coordinating
documents from the Board. They have been found to be
valuable in documenting and disseminating actions, agree-
ments, and understandings of the Board. Management review
type control charts are used to monitor progress on critical
tasks and assure all areas are up-to-date in their implementation actions.

The organization of the Fracture and Fatigue Control
Board is shown in Figure 5. This Board is the primary coordinating body for all aspects of the control plan and
fills the need for intensified interdisciplinary involvement
and coordination. It is probably the single-most important
coordination item. The Board reports directly to the Wing
Mod Program Manager who is Deputy C-5 Project Director. A
Chairman is appointed by the Wing Mod Program Manager. Representation on the Board is furnished by all disciplines and
organizations involved in critical part control. The current
representation is shown here. Additional representation is
added as necessary to ensure overall coordination of the
Plan. Some of the major functions and responsibilities of
the Control Board are:

(a) Overall coordination of the Control Plan

(b) Monitoring the control system

(c) Problem area review and recommendation

(d) Review of reports on first article inspection

(e) Direct and review control system audits

(f) Maintain close liaison with Program Manager

(g) Recommend additions, revisions, or deletions
to the Plan

Minutes are kept on all meetings. For the first eight
months it met every week. Since then, meetings have been
scheduled each two weeks. The concept of the plan management is to have the implementation tasks accomplished under
the normal supervision within each area of responsibility.
However, the Board has been a driving force and a vital and
important element in the implementation of the Plan.

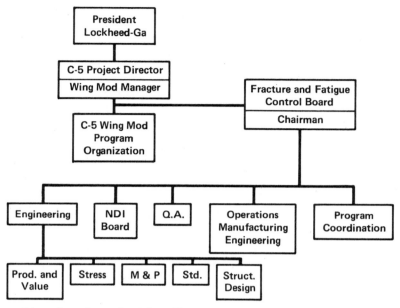

Fig. 5. Organization Chart

It is felt that the across-the-table coordination on the implementation of the Plan has allowed us to work out problems to the mutual benefit of all concerned and has resulted in a much more optimum approach than would have resulted without this person-to-person contact. It will be obvious through the remainder of the implementation discussion how closely this coordination has tied together the requirements, the inspection, and the production in order to achieve better quality.

Special Identification

Figure 6 illustrates the significant implementation actions in the area of special identification of critical parts, requirements, and controls. A Fracture and Fatigue Critical Item Memo System has been initiated. This memo is prepared by the Stress Department and distributed to all organizations with an interest in critical parts at a very early stage. The memo is basically an early warning to all concerned that a particular engineering job package contains critical parts based on preliminary evaluation, and as such it alerts Quality Assurance, Planning, and the NDI Review Board.

Three other special identification items that are all covered by a special Drafting Manual Instruction are:

(a) the critical parts list (a specific MIL-STD-1530 requirement),

(b) special drawing identification system,

and

(c) a critical requirements identification system.

The Critical Parts List is contained in a Book Form Drawing. There are 45 parts on the list, not counting left and rights. Total dash numbers counting lefts and rights add up to 79. The list is subject to change as the design progresses - either by additions or deletions. The primary purpose of a Critical Parts List is to have a reference document available for all interested organizations. However, the detail special requirements are controlled on the Engineering Drawing, not the Critical Parts List.

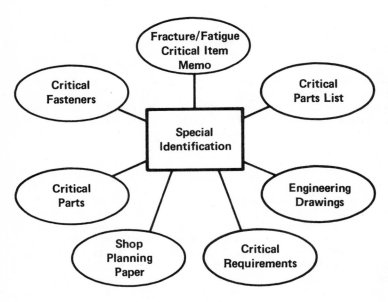

Fig. 6. Special Identification

The procedure for two other special identification items - the special shop planning identification procedure and the identification of the part itself - are covered by an authorizing directive which is a Manufacturing Engineering Procedure Document. The shop planning will have a special overstamp on all planning that covers critical parts or requirements. It will be obvious from the overstamp that some operations are of a critical or special nature and fracture and fatigue special controls are involved. In addition to the overstamp, each individual operation that requires special controls will be identified on the planning paper work. This special identification will alert the production workers, the inspector, and the liaison engineer of the requirements. The authorizing directive also requires that a special label be attached to critical parts that identifies them as fracture and fatigue critical throughout their processing. In order to allow Planning to better coordinate the fabrication operations and inspections, some of the Engineering Drawings will include an identification procedure for critical fasteners. The shop planning will then call out specific fastener inspection buy-offs using this identification.

The special drawing identification and the critical requirements identification procedures are explained by the example in Figure 7. This figure depicts an Engineering Drawing in schematic form in order to illustrate the technique being used to identify special fracture and fatigue control requirements. Directly above the title block will be overlayed "Note: This drawing contains fracture and fatigue critical requirements". This note will be clearly and unmistakably apparent. Above this note, in the list of materials, the parts that are critical will be identified by Delta Note 1. This note will read "Category (I or II) critical part per LG1-D1-S-3582".[9] (This is the Fracture and Fatigue Control Plan). To the left of the list of materials, in the general notes section, the special requirements or features will be noted by Delta Note 2 that will state "Fracture and Fatigue Critical Requirements - Verification Mandatory". This terminology identifies the requirement or feature as one that Engineering requires Quality Assurance to inspect (without delegating the inspection), and instructs that a sampling inspection is not to be used. This drawing identification procedure has many advantages but primarily improves communication of special requirements to be discussed later. It also allows the Engineer to be very selective and discrete and to apply special controls and requirements only where absolutely

necessary, not a broad-brush approach that would be unnecessarily costly.

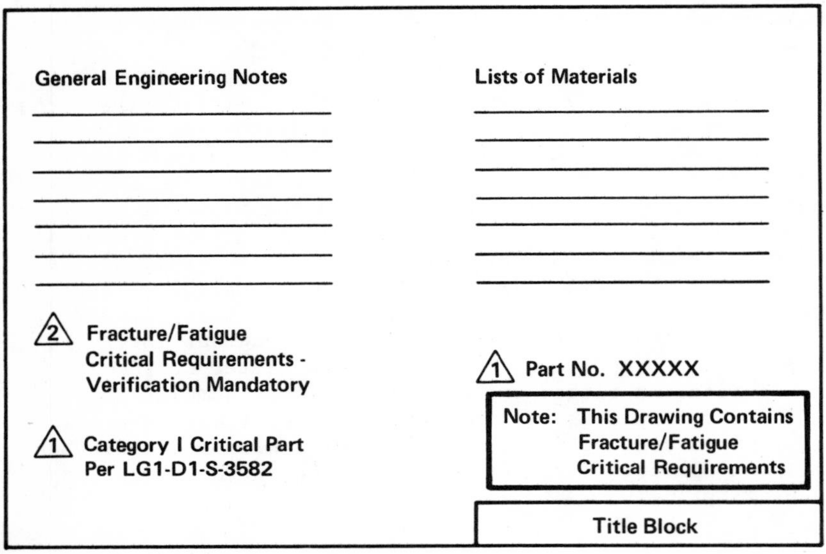

Fig. 7. Critical Drawing Identification

Special In-Process Control and Closure

Figure 8 illustrates the actions taken in the area of
special in-process control and closure. An in-plant NDI
evaluation has been made for the eddy current hole probe
and the penetrant inspection techniques. These are the
primary NDI procedures to be used during the Wing Modification Program. The Quality Assurance organization conducted a hole inspection study. The purpose of this study
was to determine inspection technique sensitivities to
various hole features such as finish, scratches, etc. The
results will allow Quality Assurance to optimize their inspection techniques in meeting Engineering requirements.
A special review of critical part and assembly shop planning
by Engineering is a significant new closure item. The purpose of this review is to help assure that the critical
operations are properly called out and accomplished in the
correct sequence and that the inspection points are well
coordinated. Shop planning changes that involve critical
feature also require this Engineering review. The special
MRB action closure item simply means that Structures Engineering will approve all MRB action where critical features

are involved in the discrepancy. This procedure is normally
followed but on critical features it will be mandatory. The
first article critical part and assembly review and report
is another new closure item that will be a combined review
by Engineering, Quality Assurance, and Production of the
final completed critical parts and assemblies to assure
that all aspects of the system are functioning as planned.
Outside production reviews and audits of the control system
will be conducted as directed by the Fracture and Fatigue
Control Board periodically to continually monitor the system.

The guidelines of making maximum use of the existing
system has obviously been followed in this area with some
significant improvements in controls and closure to give
added confidence in the overall operation.

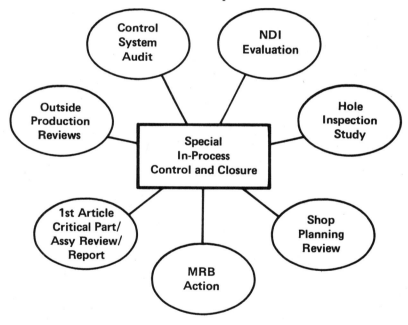

Fig. 8. Special In-Process Control and Closure

Updated Requirements

Figure 9 illustrates the significant implementation
actions in the area of updated requirements. An industry
survey and a critical fastener survey were conducted as a
preliminary to reviewing the existing requirements and

revising where necessary. The critical fastener survey
was an early assessment of: (a) locations where special
fastener installation would be a requirement, (b) tool and
inspection accessibility, and (c) potential quality problem
areas. The results of this survey played an important part
in the selection of proper fastener systems for critical
areas of the wing. New material specifications have been
developed for the special control of material that is to
be used in safety of flight structure. These include
fracture toughness guarantees and tighter controls on
impurities. Another item associated with updated requirements is the formalization of Standards Bulletins.
The bulletins outline the approved fastener systems and
proper drawing callouts to be used where these fasteners
are used in critical areas.

An existing Book Form Drawing System to outline special
inspection requirements has been adapted to the inspection
of new wing structure and wing/fuselage structure. Where
inspection requirements are considered too complex to include
on the face of the Engineering Drawing, a Book Form Drawing
will be developed to outline these requirements and called
out on the Engineering Drawing. This type of system has
been used in the past to control ECP installation on existing aircraft. A technique to allow Engineering to require a special visual inspection has been developed. The
inspection details are coordinated in formal Quality Engineering Instructions. Engineering will call out locations
for special visual inspection on the drawing and Quality
Assurance will automatically invoke the special visual
inspection instructions that include lighting, magnifica-
tion, and accept/reject criteria. The Subcontractor controls have been updated to include fracture and fatigue
control requirements. Major subcontractors are required
to provide Lockheed-Georgia Company with a plan for our
approval. Mandatory Quality Assurance verification of
critical features has been previously discussed to show
how it ties in with the callouts on the Engineering Drawing.
The agreement between Engineering and Quality Assurance on
the details of this requirement allows the engineer to require certain inspections and Quality Assurance to know
exactly what is the nature of the critical requirement.

The inspectability of the design is being given special
attention and is considered to be one of the important design
parameters. Where NDI is involved, the Inspection Book Form
Drawing is reviewed by the NDI Review Board to ensure proper
inspectability. For inspections other than NDI, a Design

Handbook revision defines the envelopes required for inspection devices. The designer in many cases now will know what
type of inspections are to be made because of the closer coordination of Engineering, Quality Assurance, and Fabrication,
and be able to plan accordingly.

Some of the most important updated requirements are contained in a new standard process control specification - the
critical fastener installation control specification. This
specification covers the installation of fasteners in critical areas and will be discussed in more detail.

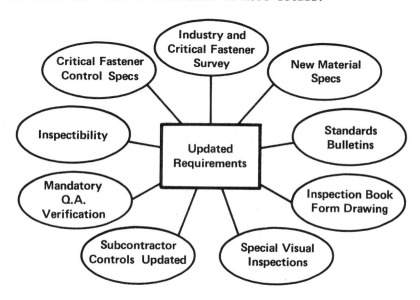

Fig. 9. Updated Requirements

Figure 10 illustrates the scope of these special
fastener control requirements.

This specification is only applicable to Category I
joints. Some joints may have their own special requirements
and be excluded from this specification. This is an overall
umbrella specification in scope. It recognizes the concept
that good quality requires properly coordinated application
of Engineering requirements, inspections, trained personnel,
and good tools. The specification applies essentially existing Engineering requirements. The critical fastener
installation specification includes minimum inspection,

training and tool control requirements. These inspection requirements are tailored to the method of fabrication and allow lower levels of inspection when the manufacturing method eliminates the possibility of the human error.

Special training is required for installers working on Category I joints. The training program is to be approved by Engineering and will include an individual certification program and annual requalification of personnel. This specification also includes tool control requirements for tools used on Category I joints. Those tools will have special identification, certification, and protection.

This specification is one of the most important new elements of the C-5A Fracture and Fatigue Control Plan and has been thoroughly coordinated, originally within the Fracture and Fatigue Control Board, then plant wide.

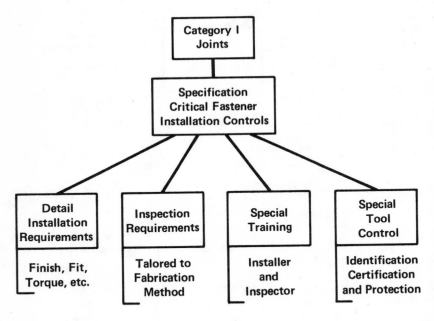

Fig. 10. Critical Fastener Installation Controls

SUMMARY

While the Fracture and Fatigue Control Plan described herein is basically done in response to Air Force requirements, there were an infinite number of ways to respond. It is felt that there have been benefits derived from the manner in which these requirements were met that are outside the normal benefits associated with satisfactory contractual performance. Among these benefits are greatly improved over-all plant wide coordination. The capability to discriminate and identify specific critical features, better coordinated planning, and the ability to better include inspectability in the design are some of the other advantages that have been recognized. And finally, the major challenge of satisfying the requirements without significant schedule or cost impact appears to have been met.

REFERENCES

(1) R. J. Gran, F. D. Orazio, Jr., P. C. Paris, G. R. Irwin, and R. Hertzborg, "Investigation and Analysis Development of Early Life Aircraft Structural Failures," U.T.C., Dayton, Ohio, AFFDL-TR-70-149, March 1971.

(2) W. T. Kirkby, "Examples of Aircraft Failure," in Fracture Mechanics of Aircraft Structures, ed. by Liebowitz, AGARD-AG-176, January 1974.

(3) "Airworthiness Standards: Transport Category Airplanes," FAR Vol. III, Part 25.

(4) H. A. Wood, "Fracture Control Procedures for Aircraft Structural Integrity," AFFDL-TR-71-89, July 1971.

(5) C. F. Tiffany and G. P. Haviland, "The USAF Structural Integrity Program," AIAA Paper 73-18, 1973.

(6) MIL-STD-1530, "Aircraft Structural Integrity Program, Airplane Requirements."

(7) MIL-A-83444 (USAF), "Airplane Damage Tolerance Requirements".

(8) MIL-I-6870C, "Inspection Program Requirements, Nondestructive Testing."

(9) Lockheed-Georgia Company LG1-D1-5-3582, "C-5A Wing Modification Program - Fracture and Fatigue Control Plan."

EVOLUTION OF FRACTURE CONTROL
IN THE B-1 PROGRAM

Mark A. Owen
Deputy for B-1
Aeronautical Systems Division

INTRODUCTION

The last ten years have seen a significant reorientation
in the USAF structural philosophy. During the time that "fly-
before-buy" was driving aircraft system procurement into long
development periods prior to production commitment, structural
requirements were also being revised. The F-111 and C-5A
programs had shown the need for an approach which recognized
the fabrication, materials selection, and inspection inputs
to structural integrity, in addition to those of the design
discipline. Structural service failures and rising inspection
and maintenance costs in older USAF aircraft had shown that
structural integrity considerations could not end with aircraft delivery, but were necessary throughout the aircraft's
useful life. Structural specifications and standards were
changed to reflect these lessons.

A USAF Aircraft Structural Integrity Program (ASIP)
document had been in existence since 1966 (1). Its provisions
included requirements for full airplane static and fatigue
tests, fatigue design using a scatter factor of four on design
life, and a fleet tracking scheme to compare actual and design
usages. The ASIP document which exists today includes both
durability and damage tolerance design, control, and testing
requirements (2). A major innovation was the addition of
fracture mechanics and fracture control concepts for use in
achieving these durability and damage tolerance requirements.

The first Air Force program in which fracture mechanics
was instituted as a design requirement was the B-1 bomber.
The B-1 development began in 1970 and has continued through
the evolution of Air Force structural philosophy. It may
therefore be considered as a test case: the various precepts

of fracture mechanics and fracture control against the hard
realities and constraints of an on-going development program.

Just as the Air Force structural philosophy was changing,
evolution naturally took place in the B-1 fracture control
process. Fracture control may be defined as integrating
design, material selection, fabrication and quality assurance
to achieve a product with minimum probability of failure during its design service life. If failure probability is to be
minimized, and long term flight safety maximized, numerous
tradeoffs may be required. Such tradeoffs came to be an
important part of the B-1 fracture control process.

This paper will review briefly the general aspects of
the B-1 fracture control system and then present some of the
specifics of the system as applied to a B-1 primary structural
component, namely, the horizontal/vertical stabilizer support
fitting. Subsequent fatigue and fracture mechanics testing of
this part pertinent to the fracture control process will also
be discussed. Finally, fracture control tradeoffs which
resulted from this testing will be outlined and some general
conclusions made about the fracture control process. The intent of this treatment is to demonstrate the evolutionary
nature of fracture control by dealing with some specific examples of interdisciplinary tradeoffs which occurred on the B-1
Program.

REQUIREMENTS

The most fundamental concept in fracture control is that
in-service cracking, whether occurring by fatigue initiation
mechanisms or developing from pre-existing defects, shall be
taken into account before the structure enters service. The
B-1 fracture control system was oriented toward this concept.
Each discipline (i.e., design, materials selection, and fabrication) contributed to a list of fracture critical parts which
would receive special attention during the development cycle.
Part selection was based on potential criticality in the
presence of service cracking, yielding a list made up of
primary structural members loaded significantly in tension,
the failure of which would cause loss of the aircraft.

The B-1 specification requirements contained provisions
for fracture control from design, inspection, and fabrication
standpoints. The design concept of fracture critical components would be either fail safe or slow crack growth to protect the structure's load carrying capability even in the
presence of manufacturing induced flaws or fatigue initiated

cracks. A nondestructive inspection demonstration program would statistically quantify the size of manufacturing defect which might be missed by the various inspection techniques to be utilized and that size would be used as a design parameter. Materials from which fracture critical parts were fabricated would be controlled by a system of procedures sufficient to preclude actual fracture toughness values inferior to those assumed in design. Analysis and testing were required to show that the development of cracks would not create a flight safety problem either prior to detection using normal field inspection techniques, or throughout the design lifetime for a part considered uninspectable in service.

Two categories of fracture critical parts were specified (3). Category I parts are those which carry a weight impact as a result of fracture mechanics even when controlled toughness specification materials are used. Category II parts would carry a weight impact with conventional specification materials, but do not possess a calculable weight increase with fracture controlled materials. In practice, however, given the fact that B-1 materials had been selected based on fracture toughness, few parts were sized by the fracture mechanics criteria. For the most part fatigue, static strength, or stiffness requirements led to actual part sizings. Nevertheless some B-1 parts, because of their unique criticality, were recognized as deserving special attention from a fracture control standpoint. A list of such parts was formulated and agreement made that regardless of any analytical weight impact such parts would be designated as fracture critical. This "exception" list was also broken down into Category I and II parts.

The designation of two part categories allowed distinctions to be made within the fracture control system. The most critical parts have the highest payoff potential for the application of comprehensive fracture control procedures. Therefore Category I procedures on the B-1 included controlled toughness material specifications, special nondestructive inspection, post-thermal process toughness verification, and material and process part traceability (4). Category II procedures included only controlled toughness material specifications and special NDI. Non-fracture critical parts were of course subject to normal quality control. On the B-1 aircraft less than fifty part drawings fell under the Category I designation and less than two hundred under Category II. There are, by comparison, thousands of part drawings making up the complete B-1 structure, illustrating the advisability, and even the necessity, of such part category distinctions.

Perhaps the major impact of the fracture control requirement on the B-1 was in material selection. For the first time
materials would be procured from suppliers with guaranteed
minimum fracture toughness values in addition to the usually
specified material properties. Fracture mechanics material
allowable data (e.g., crack growth rate, fracture toughness,
and stress corrosion threshhold) would be needed for a new set
of aircraft materials. The trend in USAF aircraft toward
higher and higher tensile strengths to reduce structural
weight fractions without accounting for the resulting material
toughness dropoff had ended.

This, in general, was the fracture control system which
became a part of the B-1 contract early in the development
phase. The importance of the design, material selection,
fabrication, and inspection disciplines to long term flight
safety was recognized, as were the details of the separate
inputs from those disciplines. As the development program
continued, new lessons were learned about integration of the
disciplines into a unified fracture control program which
would minimize structural problems in the B-1 fleet.

APPLICATION

The B-1 fracture control program may best be understood
by observing its application to a specific structural component. The evolutionary nature of fracture control may also
be observed in this fashion. The remainder of this paper will
therefore be devoted to the B-1 horizontal/vertical stabilizer
support fitting shown in Figure 1. This part is selected
because it is representative of fracture critical parts as a
class having had special considerations from all the disciplines involved in fracture control. It is a single load path
part designed as slow crack growth uninspectable. Material
selection was made on the basis of fracture toughness in addition to strength level. The part is fabricated by welding
together eight separate forgings, the major of which are shown
in Figure 2.

Even without the fracture mechanics requirement, design
of the stabilizer support fitting would have been complicated.
Although the B-1 is about the size of a 707 aircraft, it has
many fighter-like elements. The horizontal stabilizers are
all-moving as a result of full flap takeoff control surface
requirements. These same horizontals, because of greater
efficiency than spoiler-ailerons at aft wing sweep positions,
provide roll control to meet the high roll rates required
during terrain following. Figure 3 illustrates the "magnified"

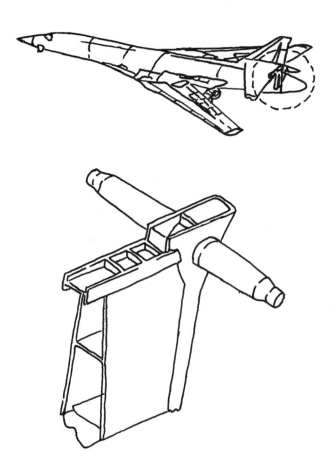

FIGURE 1.   HORIZONTAL STABILIZER SUPPORT FITTING

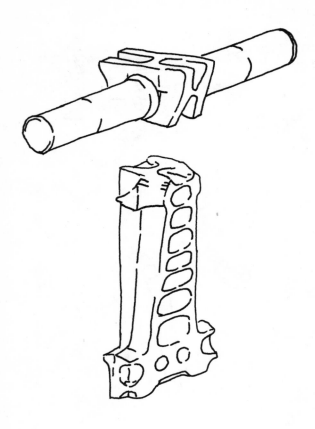

FIGURE 2.   SUPPORT FITTING BASIC FORGINGS

reacting forces in the support fitting side plates resulting
from a rolling maneuver. In theory, the vertical forces are
transfered to the side plate through the shear stresses in the
spindle arm walls. However, this transfer is very much dependent on the relative flexibilities between the spindle arm
and side plate. Instead of shearing uniformly into the side
plate, the stresses may concentrate in the vicinity of the
fillet radius. This phenomenon was observed in subsequent
testing of a complete stabilizer support fitting, but seemed
to present no particular structural life problem when the
design fatigue load spectrum for the support fitting contained
a maximum occurrence of only about 40% of design limit load.
The design condition for the majority of the fitting had been
a relatively rare heavy weight abrupt roll case in which
moments created by the left and right horizontals and the
vertical tail were all additive.

A 9Ni-4Co-.20C steel was selected for use in the support
fitting. The steel had been developed by the Navy for use in
submarine applications requiring high weldability. Its high
fracture toughness, corrosion resistance, and weldability made
it especially attractive for B-1 applications. Had there been
no fracture requirement, a higher strength steel might well
have been selected. For example, if a steel such as 300M
with an FTU of 280 ksi had been used instead of 9-4-.20
(FTU = 190 ksi) the resulting support fitting might easily
have weighed 30% less. However fracture mechanics considerations show another aspect to the problem. Critical crack
length is proportional to the square of fracture toughness.
Using fracture toughness values for 300M and 9-4-.20 of 69 and
140 ksi-in 1/2, respectively, the critical crack length of
300M may be calculated as one quarter that of 9-4-.20 at the
same applied stress level. This point will be considered
again in the section on test results.

The 9-4-.20 steel's high weldability suggested welding
as a joining process for the separate forgings that made up
the entire support fitting. A weld instead of a bolted joint
may even have partially offset weight penalties resulting from
material strength levels. The vertical location of the most
critical joint connecting the spindle arm to box beam forging
was set by manufacturing to facilitate the finish machining
process. Machining after heat treatment necessitated being
able to rotate the spindle arm forging through 360 degrees.
Machine tool availability limited the part's maximum radial
dimension, yielding a weld joint about two inches below the
spindle arm. The welding process involved preparing the joint
in the shape of a double-U groove, butting the parts together
and using a manual, multi-pass, tungsten inert gas funsion
weld with 9-4-.20 weld wire. This operation was followed by

SYMMETRIC CASE

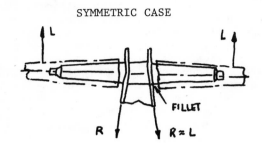

ANTISYMMETRIC CASE

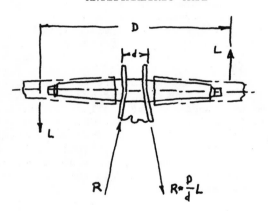

FIGURE 3. STABILIZER SUPPORT FITTING LOADS

a two hour termal stress relief at 950°F.

Nondestructive inspection techniques were utilized during the course of welding. The weld was x-rayed at the completion of the root pass, when the weld was at one-half thickness, and again at full thickness. The part was both magnetic particle and ultrasonically inspected at the completion of welding. Any anomalies uncovered were machined out and the part rewelded and re-inspected.

In all cases material procurement, fabrication, and inspection procedures were carefully documented. Fracture critical areas of parts were specifically identified on part drawings. Material and processing specifications were tightly controlled. Inspection instructions were drawn up in detail. Checks were made on fracture toughness by testing prolongations from both the basic forgings and the finished welds. In spite of these precautions, there were still some lessons to be learned as the horizontal/vertical stabilizer support fitting entered its structural test phase.

TESTING

Of all the structural areas on the B-1, two were especially critical from a structural standpoint and therefore received special attention during the structural test program. These were the wing pivot area and the horizontal/vertical stabilizer support fitting. During the design development test phase two separate support fittings were tested, one statically to design ultimate load and the other to a total of five lifetimes of the design fatigue spectrum. Two additional fittings (static and fatigue) were tested as part of the aft fuselage/empennage assembly during design verification testing. This time the fittings were tested to a revised set of loads which for the first time extended the B-1 fracture control program to the limit.

Some explaination of this loads revision is in order. The B-1 structural design loads were released in the early 1973 time period. They were derived for a B-1 which was still very much on paper. Preliminary weights and inertias had been iterated to an extent, structural influence coefficients were based on general internal sizings, and some wind tunnel data was available. However no structural dynamics had been included and only a preliminary description of the B-1's automatic stability and control augmentation system (SCAS) had been made. Using these inputs the maximum load in the support fitting fatigue spectrum was 40% of design limit.

As the time approached to begin fatigue cycling the design verification test articles, more information had been obtained. Revised mass distributions and inertias had been calculated, along with updated influence coefficients. Structural dynamic effects were available for the vertical and lateral axes. Aerodynamic definition had been refined such that vertical and horizontal tail interference effects were predicted. Finally as a result of information gained from the full flight control simulator, the importance of the SCAS to the loading spectrum on the horizontal tail was becoming clear. A decision was therefore made to test the design verification test articles to a spectrum of loads derived using these latest inputs rather than hold to the design spectrum. The revised spectrum contained a maximum load of approximately 70% of design limit.

This structural test decision was proven sound when the first B-1 entered flight test. The early flight test program was unusually severe because of the need to test the B-1's handling qualities throughout the flight envelope. Even so, the days of a mild fatigue spectrum for the stabilizer support fitting were gone forever as shown by a comparison of a "flight" from the design, updated, and flight test spectra in Figure 4. In view of the difference between the test and design spectra, some fatigue cracking during the aft fuselage design verification test would not have been surprising. When the first crack was found in the fracture critical stabilizer support fitting, it nevertheless had considerable significance to the B-1 production design and to the fracture control program in general.

At about three-quarters of one fatigue lifetime, a crack was discovered in the stabilizer support fitting side plate weld joint. The crack was approximately five inches long and completely through the side plate thickness in the vicinity of a weld relief hole in the centerline web. (Figure 5 is a section cut through the spindle arm to box beam joint showing geometrical details.) The fracture faces were removed for laboratory analysis which revealed that the crack had initiated from a small (0.02x0.04 inch) defect-like zone in the weld on the inner surface of the side plate (5). Analysis further showed that the propagation rate through the side plate was significantly greater than would have been predicted given the stress field at this location, but, at five inches long, the crack had not reached critical length. (By comparison the 300M referred to earlier would have failed catastrophically before this particular crack was detected. In this case the fracture toughness of the 9-4-.20 steel was worth any associated weight penalty).

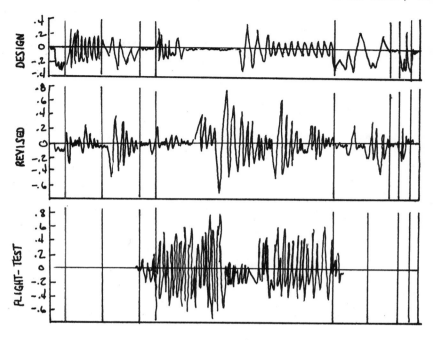

FIGURE 4. LOAD SPECTRA COMPARISON

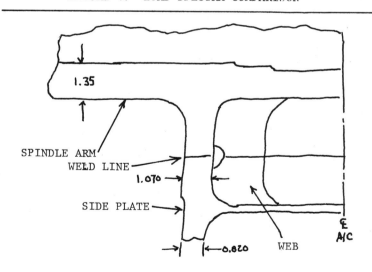

FIGURE 5. FILLET AND SIDE PLATE AREA

The crack's detection precipitated an extensive review of the stabilizer support fitting. One of the first findings was the existence of the stress buildup, or concentration, in the vicinity of the fillet radius referred to earlier. Later X ray diffraction and small element tests revealed that, due to the thickness of the sideplates, fairly significant residual stresses were likely to remain in the weld after thermal stress relief. These findings served to explain the crack growth rate observed on the fracture face. Figure 6 shows crack growth precictions for the side plate weld joint with and without accounting for weld residual stresses (6). The data point from this test is also shown.

Following repair of the cracked area, testing of the aft fuselage article was resumed with periodic nondestructive inspections of the remaining side plate. Eventually a crack was found by dye penetrant inspection which had initiated from an internal weld defect and propagated to the surface. Ultrasonic inspections had been unsuccessful in finding the crack prior to surface break-through. However subsequent failure analysis revealed, not a clean well-defined fatigue crack, but a series of multiple origins along the defect propagating on slightly different planes. This may have hopelessly complicated the ultrasonic signals.

An attempt was made to better understand the detectability of fatigue cracks initiating from subsurface weld defects by testing small welded coupons. Internal artificial flaws were cycled to the stabilizer support fitting fatigue spectrum. In these tests internal crack growth was successfully monitored by ultrasonic inspection. However failure analysis revealed that the coupon fatigue striations were well defined and in the same plane whereas growth in the stabilizer side plate had been along different planes from multiple origins. As a result, no conclusion could be made about the detectability of an internal fatigue crack in the stabilizer support fitting. On the other hand, testing had revealed that any internal fatigue crack would reach the surface well before it reached critical length.

These design verification test cracks required each element of the fracture control discipline to re-evaluate its inputs to the stabilizer support fitting configuration in the light of a more severe usage spectrum. The next section of this paper will discuss results of this evaluation and some of the fracture control tradeoffs made as a result of the lessons learned during structural testing.

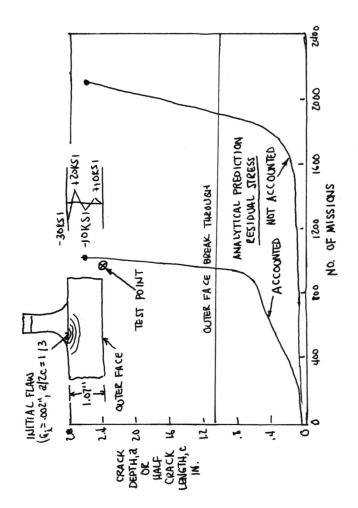

FIGURE 6. SUPPORT FITTING SIDEPLATE CRACK GROWTH

EVALUATION AND TRADEOFF

The B-1 design verification test program had been scheduled so that testing would be completed two full years ahead
of delivery of the first production aircraft. Consequently
time was available to take advantage of lessons learned by
incorporating changes into the production horizontal/vertical
stabilizer support fitting. The most important task was to
understand the results obtained in order to make changes which
would minimize the occurrence of similar cracking in the
future B-1 fleet.

Several general observations could be made about the
fracture control program on the stabilizer support fitting.
First, the design development test fitting which completed
five lifetimes of the design spectrum, including more than one
lifetime of crack growth without failure, demonstrated the
feasibility of the basic design concept, the material selected,
and the use of welds for joining in a moderate stress field.
If the weld joint were located in a moderate stress field
relative to the more severe loading spectrum, long life might
again be expected. Second, each side plate crack initiated
in the same specific location - the vicinity of a weld relief
hole in the centerline web of the support fitting. This area
is more difficult to weld because of accessibility problems
created by the web. Improving accessibility might mean
improving the overall weld quality. Finally, each crack initiated from a pre-existing defect zone in the weld area that
had apparently been missed by nondestructive inspection. This
aspect needed more attention prior to establishing a final
production configuration for the stabilizer support fitting.

During the nondestructive inspection demonstration program, little subsurface crack detection work had been done.
The program had been oriented toward the "most critical" crack
which was assumed to be a surface flaw based on stress intensity considerations. (The assumed stress intensity for a
surface crack is 10% higher than an internal crack for the
same geometry and stress level (7).) In fact the growth rate
of the surface crack found at 0.75 lifetimes on one side of
the design verification test article was faster than the
internally initiated crack found at 2.3 lifetimes on the other
side. The smallest surface flaw demonstrated as 90% detectable with 95% confidence level was 0.050 inch in depth. Consequently the 0.02x0.04 inch defect mentioned earlier might
well have been missed by quality control inspections. Based
on these observations, a conservative 0.05 inch surface crack

could still be used to design the production support fitting.

Tradeoffs were clearly in order between the various disciplines concerned with fracture control. To get the weld joint into a lower stress field, it needed to be moved out of the vicinity of the spindle arm. A tradeoff was made between design and manufacturing to locate the joint eleven inches lower on the box beam forging, cutting the stress level in half. This necessitated new forging dyes and locating a tool with large enough radius about the rotational axis to finish machine the spindle arm.

In an effort to improve the weld quality and decrease any stress concentration contributed by the presence of the weld relief hole, the weld configuration was changed to a single-U instead of double-U groove. This allowed welding to be done from the outside of the side plate in one continuous pass instead of laying weld material through the weld relief hole. The effect of these changes was to make the weld more tolerant of residual stress remaining after stress relief.

From an overall program standpoint, these changes do not seem particularly significant. Moving a weld joint location and altering a weld procedure did not represent a major redesign or structural concept change. Some consideration in fact was given to major redesign, but the cost and weight penalties would have been significant. Such major ramifications were saved by merely integrating the fracture control system.

This example has shown the interdependence of design, fabrication, and inspection. The location of a weld joint, or any joint, may not be left entirely to manufacturing. Design analysis must be updated to take into account perturbations created by the manufacturing process. Quality control inspections may leave "acceptable" defects which will affect the life if stress levels or loading spectra are higher than predicted. Fracture control is a process in which design, fabrication, material selection, and inspection impact one another, requiring the total process to be continually updated to reflect new inputs or new requirements.

CONCLUSIONS

This treatment of the aircraft fracture control problem has shown by a specific example the role that historical aircraft disciplines such as design, materials selection, fabrication, and inspection have in the relatively new discipline of fracture control. These disciplines are not indpendent

but interrelated in the fracture control process. In order
to achieve the goal of minimum probability of failure in
service, tradeoffs between disciplines will be required
since basic tenets may sometimes be at cross purposes.

Fracture control has also been shown to be an evolutionary process. Early procedures, instructions, and specifications can only cover "known" trouble spots. Aircraft development has historically been full of unknown or unexpected problem areas. The fracture control process must be flexible
enough to re-evaluate itself when new inputs arise. The
result will be a unified discipline which can react to all
manner of aircraft design constraints.

The impact of fracture control cannot necessarily be
assessed in any quantitative way. Time will tell whether or
not aircraft such as the B-1 which utilized fracture control
concepts during development will exhibit fewer structural
problems in service than past aircraft. About all that can
be said at this stage is that the fracture control program
has brought the attainment of a quality product to a status
where it is left less to chance than to the integration of
sound design, fabrication, material selection, and inspection
practices.

REFERENCES

(1) Wells, Harold M., and King, Troy T., "Air Force Aircraft Structural Integrity Program: Airplane Requirements," ASD-TR-66-57, May 1970.

(2) Anon., "Aircraft Structural Integrity Program, Airplane Requirements," MIL-STD-1530A, 11 December 1975.

(3) Anon., "Memorandum of Agreement, ECP B1-0092 (Fracture Mechanics)" Rockwell International, NA-71-958, July 1971.

(4) Padian, W.D., "B-1 Fracture Control System," Rockwell International, NA-72-383-1, June 1972.

(5) Young, J., "Failure Analysis of DVT-2 Aft Fuselage Horizontal/Vertical Stabilizer Support Fitting Crack at Right Hand Weld Relief Hole," Rockwell International, TFD-76-1086, April 1977.

(6) Anon., "Briefing to Scientific Advisory Board Committee on B-1 Structures," Rockwell International, NA-77-280, March 1977.

(7) Szamossi, M., "Crack Propagation Analysis by G. Vroman's Model," Rockwell International, NA-72-94, February 1972.

LIQUEFIED GAS SHIPS
DESIGN AND CONSTRUCTION REQUIREMENTS
FOR CARGO TANKS AND HULL

T. C. GREENE

and

P. J. PLUTA

Merchant Marine Technical Division
U. S. Coast Guard Headquarters
U. S. Department of Transportation

A paper for presentation at the ASNT/ASM Fifth Annual
Forum, June 1977, at Tarpon Springs, Florida

ABSTRACT

On October 4, 1976, the U. S. Coast Guard published a
set of Proposed Standards for Self-Propelled Vessels Carrying
Bulk Liquefied Gases. These standards implement the corresponding international code developed under the auspices of
the International Maritime Consultative Organization (IMCO),
formally, IMCO Resolution A.328(IX), the Code for the
Construction and Equipment of Ships Carrying Liquefied Gases
in Bulk.

This Code, which is applicable to new ships, was developed
to provide an international standard for the safe carriage by
sea in bulk of liquefied gases. It contains design and construction requirements for ships involved in such carriage and
the equipment they should carry so as to minimize the risk to
the ship, its crew and to the environment. The proposed standards use the IMCO Gas Code as a base document and, when published as a Final Rule, will apply to all new U. S. flag
liquefied gas ships and those new foreign flag gas ships which
call on U. S. ports. Existing vessels and vessels under construction during development of the regulations are given
separate treatment.

This paper reviews the role of the Coast Guard and IMCO
in maritime safety, offers a brief look at the industry as it
exists today, and focuses on those portions of the proposed
standards which address the special design and construction
requirements for the cargo tanks and ship's hull.

INTRODUCTION

The Coast Guard is responsible for the development and
enforcement of standards for the design, construction, alteration, repair, maintenance and operation of merchant vessels
documented under the laws of the United States and foreign
merchant vessels carrying liquefied gases in bulk in navigable
waters of the United States. To carry out this responsibility,
the Coast Guard reviews the properties of the cargoes carried
and identifies the associated hazards. It then specifies the
necessary standards for a particular cargo or type of vessel
to ensure that the associated hazards are properly addressed.
This is done by promulgating regulations which govern the
design, construction, inspection and operation of the vessel,
reviewing and approving the construction plans, inspecting
the vessel to verify that it is built in accordance with the
approved plans, and exercising operational controls relating
to the vessel and the ports that it intends to enter. This

combination of design, construction and operational requirements is intended to provide for the safety of life, property,
and the marine environment in U. S. ports regardless of the
flag the vessels flies. (1)   In keeping with the process
described above, the U. S. Coast Guard recently published a
set of Proposed Standards for Self-Propelled Vessels Carrying
Bulk Liquefied Gases. These Proposed Standards implement the
corresponding international code developed under the auspices
of the International Maritime Consultative Organization (IMCO);
formally, IMCO Resolution A.328(IX), the Code for the
Construction and Equipment of Ships Carrying Liquefied Gases
in Bulk, the IMCO Gas Code.

This parent document was generated in the broad international forum of IMCO. This organization was formally
established in 1958 with the cooperation of the governments of
many countries in order to maintain safe and efficient standards for shipping cargoes across international waters. (2)
Although multinational diplomatic conferences had previously
been convened to deal with specific maritime problems, the
complex organizational and administrative responsibilities fell
upon the host (convening) nation in lieu of a static, representative body. When Japan deposited the required twenty-first
ratification (at least seven of the twenty-one were required
to be from countries with at least one million gross tons of
shipping each) on March 17, 1958, the Convention ("treaty") on
the Intergovernmental Maritime Consultative Organization was
officially established.

The many technological, environmental, and political
factors affecting maritime commerce require the continual
review of past conventions and resolutions and the drafting
of new ones. IMCO provides a forum where the maritime regulatory agencies of its member-goverments can consult together
to accomplish this work. IMCO's functions encompass three
broad categories: promotion of safety at sea and efficiency
of navigation, prevention of marine pollution from ships and
other craft, and other work relating to shipping activities.
These functions are carried out through participation in IMCO-
sponsored diplomatic conferences, special meetings, and sym-
posiums which ultimately results in the development of a convention or resolution. These documents are recommendations
to national governments which reflect the consensus of the
participating agencies, and their effectiveness depends entirely upon the cooperation of the signatory countries. IMCO
has no enforcement powers other than the implied power of the
consensus and approval of its member-states.

As evidenced in Figure 1, IMCO's main bodies consist of the Assembly, the Council, the Maritime Safety Committee, the Marine Environment Protection Committee, and the Secretariat. The Assembly is the policy-making body of IMCO, and its membership consists of representatives from the governments of all member-states. The Assembly ordinarily meets in London once every two years.

The Council is IMCO's governing body between Assembly sessions and meets twice a year. The Council members are government representatives from 18 member-states who are elected by the Assembly for terms of 2 years. Council members are those nations which have the greatest trade and shipping interests.

The work of IMCO relating to safety is carried out under the direction of the Maritime Safety Committee which consists of representatives of 16 member-governments who are elected by and from the Assembly for a term of 4 years. At least 8 of the 16 representatives elected must be from 10 of the largest shipowning nations. Of the remaining 8 representatives, 4 must be elected to assure broad geographic representation. Because of the multiplicity and complexity of technical problems, the Maritime Safety Committee has organized subcommittees and groups of experts, which consist of representatives from any IMCO member-state. One of these, the Subcommittee on Ship Design and Equipment, drafted the IMCO Gas Code.

The Marine Environment Protection Committee, open to all members of IMCO, is responsible for coordinating IMCO's activities in the prevention and control of marine pollution caused by ships.

The Secretariat of IMCO disseminates a great quantity of published information to members and other organizations. The Secretariat consists of the Secretary-General, the Deputy Secretary-General, the Secretary of the Maritime Safety Committee, and a staff of 175 international civil servants recruited on a wide geographical basis. IMCO is headquartered in London, England, and is supported by contributions from member-governments which are based on assessments according to tonnage by flag.

Subject matter for possible adoption as an international standard normally originates at the Committee level. The cognizant Committee studies the problem and proposes a recommended course of action to the Assembly via the Council. The Assembly has final authority on its approval and funding. Once

approved, the problem area is studied by a Subcommittee,
which then submits its findings to the Committee. If adopted,
these findings are forwarded to the Council and, finally, the
Assembly for approval. If the Assembly approves, the item
becomes a recommendation by IMCO to its member-states for
adoption. It is then incumbent upon the individual governments to enact national legislation or regulations to make
the recommendation applicable to their own ships or waters.

The immediate case, the IMCO Gas Code, was prepared by
an Ad Hoc Working Group of the IMCO Subcommittee on Ship
Design and Equipment. The Group was composed of delegations
from Belgium, Canada, Denmark, Finland, France, Federal
Republic of Germany, Italy, Japan, Liberia, Netherlands,
Norway, Poland, Sweden, Union of Soviet Socialist Republics,
United Kingdom and the United States. (3)   Observers from a
number of international technical and trade organizations also
participated in the work of the Group. Individual delegations
regularly contained governmental as well as leading industrial
experts on gas ship design, construction and operation. The
broad representation and the size of the Working Group are
good measures of the importance with which IMCO viewed the
work. Moveover, it should be noted that the efforts of this
Working Group have been recognized through the upgrading of
its position to full subcommittee stature, as the Subcommittee
on Bulk Chemicals.

In order to more effectively participate in the development of the Code, the Coast Guard formed a special task
group under the Chemical Transportation Industry Advisory
Committee. Members of this task group represented affected
industry groups and associated engineering disciplines. The
task group met frequently with the Coast Guard to develop and
review the emerging Code, and several of its members participated along with Coast Guard personnel at IMCO meetings.

Through the efforts of all concerned, the Ninth Assembly
of IMCO adopted IMCO Resolution A.328(IX), the IMCO Gas Code,
in November 1975.

More recently, the Coast Guard has taken action to implement the IMCO Gas Code into U. S. regulations. These standards were published as Proposed Rules in October 1976 for
public review and comment. Comments have been received and
considered, and associated revisions to the regulations are
currently being made. When the changes are administratively
cleared within the Department of Transportation, the Proposed
Standards will be published as a Final Rule, binding on all

new U. S. flag liquefied gas ships and those new foreign flag gas ships which call on U. S. ports. Existing vessels and vessels under construction during development of the regulations are given separate treatment.

In developing these standards, as in developing the IMCO Gas Code, the major concern was the possibility of a cargo release posing a hazard to a wide area. While most liquefied gases do not pose a water pollution threat, other potential hazards such as flammability, toxicity and the extreme low temperature of carriage call for special attention to cargo containment under both normal and emergency conditions. The central theme or philosophy is to provide maximum attention to cargo containment and to minimize the release of cargo in the event of a casualty. The regulations have, therefore, been based upon sound naval architecture and engineering principles and the best understanding of gas ship technology today. They emphasize ship design and equipment; but recognize that, in order to ensure the safe transport of liquefied gases, the total system must be appraised. Other equally important facets of the system, such as operations, traffic control, and handling in port remain primarily the responsibility of the individual governments where the vessels trade.

The regulations contain a wide variety of detailed requirements governing the design and equipment of new ships. The remainder of this paper will focus on the standards for two of the more critical features of liquefied gas ship design, the cargo containment system and the ship's hull. Treatment of these areas will be slanted toward their impact on vessel design through materials selection, welding, and nondestructive testing.

CARGO TANKS

Tank Types

At the heart of every liquefied gas ship are its cargo tanks, the primary elements of the cargo containment system. Each cargo containment system must safely contain the cargo under all normal operating and several possible emergency or damaged conditions. In the event of a cargo release, the flammability and extreme low temperature (as low as $-260^{\circ}F$) of many cargoes, plus the toxicity of some would pose a serious hazard to the ship and the ship's crew. Even more important, in the event of a large cargo release, a wide area surrounding the ship could also be exposed. One of several means available to minimize the risk is by preventing accidental cargo release by careful attention in design and construction of the cargo containment system.

To achieve this, the containment system should be designed to perform the following major functions:

a. Contain the cargo under all normal operating conditions;

b. Protect the ship's hull from the low temperature of the cargo to prevent brittle fracture of the hull steel;

c. Safely contain released cargo caused by minor leakage from or complete failure of the primary cargo container; and

d. Allow for early detection of system deterioration or cargo leakage so that corrective action may be taken before catastrophic failure occurs.

To perform all its functions, a containment system is required to have several distinct parts or features, some of which may be eliminated if other parts perform the required function or are sufficiently reliable to preclude a requirement for redundancy. The features required in most containment systems are:

1. A primary barrier which is the inner element of the system designed to contain the cargo when the cargo containment system includes more than one boundary.

2. A secondary barrier which is the outer element of a containment system designed to contain any envisioned leakage of liquid cargo through the primary barrier for a period of 15 days and to prevent the cooling of the ship's hull structure to an unsafe temperature.

3. Insulation which protects the ship's hull structure from the low temperature of the cargo and reduces the heat flow into the cargo limiting the cargo boil-off to an acceptable level for safety and economic considerations.

Additional systems necessary to work in conjunction with the containment system are:

1. A gas detection system which monitors spaces outside

the primary barrier to detect leakage of cargo vapor providing
an early warning of possible containment system failure and
an indication of any accumulation of dangerous flammable
vapors.

2. A temperature monitoring system to measure the
temperature of containment system parts and hull steel
adjacent to the containment system to warn of abnormally low
temperatures, resulting from primary barrier failure or
insulation breakdown, which could eventually lead to failure
of the hull structure.

Many different types of containment systems have been or
are under development throughout the world. The current types
of cargo containment systems that have been defined are summarized as follows: (3)(4)

a. Integral tanks form a structural part of the ship's
hull and are influenced in the same manner and by the same
loads which stress the adjacent hull structure.

b. Membrane tanks are non-self-supporting tanks which
consist of a thin layer (membrane) supported through insulation by the adjacent hull structure. The membrane is designed
in such a way that thermal and other expansion or contraction
is accommodated without unduly stressing the membrane.

c. Semi-membrane tanks are self-supporting tanks when
empty and consist of flat surfaces connected by shaped corners
that can expand and contract due to thermal, hydrostatic and
pressure loadings. Their design and construction may follow
the requirements for membrane tanks or independent tanks type
A and B. When loaded, the top, bottom, and sides must be
supported through insulation by the adjacent hull structures.

d. Independent tanks are self-supporting; they do not
form part of the ship's hull and are not essential for hull
strength. The three categories of independent tanks are
described below:

1. An Independent Tank Type A is usually of pris-
matic shape and designed using classical structural analysis
procedures to ship classification society rules.

2. An Independent Tank Type B employs the "leak
before failure" approach to design by using model tests,
refined analytical tools, and analysis methods to determine

stress levels, fatigue life, and crack propagation characteristics. With the exception of those semi-membrane tanks designed to type B independent tank standards, tanks of this type are usually designed as unstiffened spherical shells.

3. An Independent Tank Type C is a tank designed as a pressure vessel where the dominant stress producing load is design vapor pressure.

In a number of instances, a secondary barrier which acts as a temporary containment for an envisaged leakage of liquid cargo through the primary barrier must be provided.

The proposed Coast Guard requirements for a secondary barrier in relation to tank type and cargo temperature are shown in Table 1.

Table 1 – Secondary Barriers For Tanks (4)(5)

Tank type	Cargo Temperature At Atmospheric Pressure		
	-10°C and warmer	Colder than -10°C -55°C	Colder than -55°C
Integral - - -	No Secondary barrier required	Tank type not usually allowed	Tank type not allowed
Membrane ------------------------		Complete secondary barrier[1]	Complete secondary barrier[2]
Semi-membrane --------------------		Complete secondary barrier[1]	Complete secondary barrier[2]
Independent			
Type A----------------------		Complete secondary barrier[1]	Complete secondary barrier[2]
Type B----------------------		Partial secondary barrier[1]	Partial secondary barrier[2]
Type C----------------------		No secondary barrier required	No secondary barrier required

[1]The hull may be used as a secondary barrier.
[2]A separate secondary barrier is required.

The purpose of the secondary barrier is to contain the amount of cargo assumed to have leaked through a failed primary tank and to protect the ship's hull from being cooled below a safe temperature. The need for a secondary barrier on the different tank types was developed by IMCO and was based on the following assumptions discussed by Kime, et al. (3):

1. Integral tanks with a design temperature below -10°C, membrane, certain semi-membrane, and independent tanks type A are designed, fabricated and tested only in general conformity to ship classification society standards for tanks. These tanks, because of their general shape and arrangement, are not stress determinant considering the design methods and allowed construction tolerances. Therefore, complete failure of the primary tank barrier must be assumed as a design condition and a redundant container must be provided in the form of a complete secondary barrier.

2. Independent tanks type B and certain semi-membrane tanks, designed using the same methods as a type B, are better suited for stress analysis because of their simpler construction (usually unstiffened shells) and tight construction tolerances. These tanks are fabricated from materials that have been thoroughly tested and that have demonstrated high resistance to crack initiation and propagation. However, because the design loads are highly theoretical, the possibility of a leak sometime in the life of ship must be considered. Therefore, these tanks are required to be fitted with a partial secondary barrier or "drip tray", which will accomodate the leakage through a predicted size crack. These tanks are said to exhibit "leak before failure" characteristics because any through crack would result in detectable leakage long before the crack grew to a length that would cause catastrophic failure of the tank.

3. Independent tanks type C are not required to have a secondary barrier. These tanks, which are pressure vessels and built to code standards have a long history of outstanding service experience and are sufficiently reliable to preclude a requirement for a secondary barrier.

4. Any tank designed for cargo temperatures of -10°C or warmer is not required to have a secondary barrier to protect the hull.

5. Because of the thermal stress considerations, the ship's hull structure may not act as the secondary barrier for cargo temperatures colder than -55°C.

Tank Materials

In the selection of material for use in liquefied gas cargo tanks, the material property that immediately comes to mind is toughness at the cargo temperature. Design cargo temperatures can range anywhere from as cold as -196°C for liquid nitrogen to -165°C for liquefied natural gas (LNG), -55°C for propane, and above 0°C for vinyl chloride.

A great deal of research has been conducted to investigate the mechanical (especially fracture toughness) properties of a variety of materials for use as primary and secondary barriers of cargo containment systems. The major emphasis has been placed on metallic materials, many of which have been commonly used in low temperature or cryogenic service materials such as polyurethane foam and fiberglass reinforced plastic have been or are being investigated; however, the discussion here will be limited to the metallic materials.

Existing Coast Guard regulations identify acceptable ferritic, heat treated ferritic, and high alloy steels and make provision for the acceptance of aluminum alloys for the construction of pressure vessel and other type cargo tanks for low temperature (below -18°C) service. (6) From these existing requirements and recent experience, an expanded list of acceptable materials was developed and became part of the accepted materials in the IMCO Gas Code, the provisions of which the Coast Guard has incorporated in its new Proposed Standards. A list of currently accepted materials and their corresponding minimum service temperatures is summarized in Table 2 and Table 3.

The Coast Guard bases acceptance of material toughness on Charpy V-notch impact tests per ASTM Standard E-23. Both base material and welds must meet the minimum impact energy requirements listed in Table 4 for transverse oriented Charpy impact specimens and must be tested at a temperature 5°C (10°F) or lower below the design temperature. For temperatures of -196°C or lower, tests may be conducted at the design temperature. For weld procedure qualifications, three transverse oriented Charpy V-notch specimens must be taken from each of five locations as shown in Figure 2 and described as follows:

TABLE 2   CARGO TANK MATERIALS (Warmer than 0°C Down to −55°C)

Design Temperature	Minimum Material Requirements		
0°C or Warmer	C−Mn Steel	o	Fine Grain
		o	Normalized or Q and T
0°C Down to −55°C	C−Mn Steel	o	Fully Killed
		o	Fine Grain
		o	0.16%C Maximum
		o	Normalized or Q and T

TABLE 3   CARGO TANK MATERIALS (Colder than −55°C down to −165°C)

Minimum Design Temperature, °C	Minimum Material Requirements
−60	o 1 ½% Nickel Steel (Normalized)
−65	o 2 ¼% Nickel Steel[1]
−105	o 5% Nickel Steel[1]
−165	o 9% Nickel Steel[2]
	o Austenitic Stainless Steel (Types 304, 304L, 316, 316L, 321, 327, 347)
	o Aluminum Alloy (5083−0)
	o 36% Nickel Steel

[1]Normalized or Normalized and Tempered
[2]Double Normalized and Tempered or Q and T

TABLE 4   CHARPY V−NOTCH IMPACT REQUIREMENTS

Size of Specimen	Minimum impact value required for average of each set of 3 specimens, foot-pounds[1]	Minimum impact value permitted on one specimen only of a set, foot-pounds
10 x 10 mm ------------	20.0	13.5
10 x 7.5 mm -----------	16.5	11.0
10 x 5 mm -------------	13.5	9.0
10 x 2.5 mm -----------	10.0	6.5

[1]Straight line interpolation for intermediate values is permitted.

1.   Three specimens with the notch centered in the weld metal.

2.   Three specimens with the notch centered on the fusion line.

3.   Three specimens at each of three locations with the notch centered in the heat affected zone as follows:

  a.   1 mm from the fusion line

  b.   3 mm from the fusion line

  c.   5 mm from the fusion line

Drop weight tests per ASTM Standard E-208 are also accepted for qualification of base material and production weldments in certain cases. Drop weight tests are used in the case of materials for which the acceptable Charpy V-notch values of Table 4 can be shown to be inapplicable or for a retest in qualification of base material or production weldments which have failed a Charpy test and retest. For a drop weight test two specimens are to be tested and must exhibit a no-break performance at a test temperature at least 5°C below the design temperature.

In addition to meeting the toughness requirements already discussed, materials used in construction of independent tanks type B must be subjected to extensive fracture mechanics analysis to determine critical crack lengths and crack growth rates. For each application it is necessary to establish the following:

1.   A surface flaw will grow in depth to a through-thickness crack before it grows to a critical (unstable) crack length.

2.   An initial through-thickness crack will not grow to critical crack length when subjected to a dynamic load distribution representing the most severe 15 day period predicted during the life of the ship.

Establishing the above will ensure the ability to detect an initial through crack by detecting the cargo that would escape, thus providing at least a 15 day warning before catastrophic failure of the tank.

Fifteen days was set by the Coast Guard as a reasonable time period to allow a ship to proceed to a port where it could off-load cargo from a leaking tank.

The materials tested for use in independent tanks type B, notably 5083-0 aluminum alloy and 9% nickel steel, have met the fracture toughness requirements with considerable margin as extensively reported in the literature. (7)(8)(9)

A design vapor pressure (not less than the maximum allowable relief valve setting) limitation of 0.25 kp/cm^2 (3.55 psig) is placed on integral, membrane, semi-membrane, and independent type A tanks. With special consideration, an increase to 0.7 kp/cm^2 (10 psig) can be permitted where tank scantlings and insulation are modified as appropriate. Independent tanks type B constructed primarily of flat surfaces, are limited to 0.7 kp/cm^2 (10 psig). For other independent tanks type B the limit is subject to special consideration for each application. The pressure limit for independent tanks type C is a function of design primary membrane stress, allowable dynamic membrane stress, tank dimensions and specific gravity of the cargo, a complex issue discussed by Kime, et al. (3)

Tank Nondestructive Testing

In reviewing the cargo containment requirements described thus far, it becomes apparent that the cargo tank materials ultimately selected for construction and the procedures for making the tank shell and attachment welds are tightly controlled in order to provide the pre-construction conditions most conducive to successful construction and operation. The system loop, however, would not be complete without some means of feedback; namely, nondestructive testing. The extent and type of testing are dependent primarily upon the tank type, and focus on the verification of tank welds, leak tightness, and strength. The individual testing programs proposed by the Coast Guard are enumerated below by tank type: (4)

Membrane Tanks. The individual nature of membrane tank designs often leads to a "special consideration" situation. Nondestructive testing is no exception. No hard and fast rule can be applied to shell weld testing, but radiographic and dye penetrant testing has been employed where possible in existing designs. Also, a periodic check on production welding, commonly referred to as production testing, is conducted when applicable.

One of the more meaningful tests for this tank type is the leak tightness test. The proposed standards require a soap bubble test, a vacuum box test, or another test specifically approved for this purpose. Tests commonly used for membrane tanks are a halogen or ammonia leak detection test, which employs either a portable halogen "sniffer" or an ammonia indicator to detect leaks at the seams on one side or the membrane while the other side is pressurized with the detection gas. This test, as with the leak tightness tests for all of the tank types, investigates 100 percent of each tank shell seam.

Finally, the strength of each completed membrane tank and any space adjacent to the hull structure that supports the membrane and contains liquid is nondestructively tested using a hydrostatic, hydropneumatic, or pneumatic test, conducted in accordance with a procedure specifically approved for the immediate application.

Semi-membrane Tanks. Semi-membrane tanks with design temperatures of -20°C or colder are subject to radiographic examination of 100 percent of each full penetration butt weld in the tank shell. For those semi-membrane tanks with warmer design temperatures, radiographic testing of only shell weld intersections and 10 percent of the remaining shell welds is required. In either case, the tank welds other than shell welds must be tested using either the magnetic particle or dye penetrant method. Production testing, however, is required only when the design temperature is -18°C or colder, but must include tensile, bend, and Charpy tests for each 50 meters of full penetration butt weld on the shell.

The options on leak tightness testing are the same as for membrane tanks, but the type of test selected varies with each tank design.

The requirements for testing the strength of each semi-membrane tank are the same as those enumerated above for membrane tanks.

Independent Tanks - Type A. Shell welds, welds other than shell welds, and production testing for type A independent tanks must meet the same criteria specified for semi-membrane tanks.

Requirements for leak tightness testing, although identical to those for all other tank types, are usually met through a combination of soap bubble and vacuum box testing.

The free-standing nature of a type A independent tank, coupled with its inherent design philosophy, dictates a hydrostatic or hydropneumatic strength test which approximates the tank's service conditions as closely as possible. For a cargo tank designed to carry LNG, approximately half as dense as water, a hydrostatic test would likely result in overstressing of the tank structure, and, for that reason is not normally conducted. But partial filling of the tank with water and the addition of an air head, a hydropneumatic test, provides an acceptable and safe alternative.

Independent Tanks - Type B. The overall nondestructive testing program for type B independent tanks is nearly identical to that for type A tanks. The major difference, however, is that regardless of design temperature, 100 percent of each full penetration butt weld in the tank shell must pass a radiographic test and an ultrasonic test. The latter method was included to complement the radiographic testing, primarily in those areas where thicker sections are employed. With the emphasis on section thickness, it is likely that the Final Rule will call out ultrasonic testing only for the areas of greater thickness.

Independent Tanks - Type C The requirements for type C independent tanks, pressure vessel tanks, are patterned after existing codes for these vessels. Coast Guard regulations in Part 54 of Title 46 of the Code of Federal Regulations adequately address most general pressure vessel requirements, but additional considerations based on the application were needed. (6)

Radiographic testing of 100 percent of each full penetration butt weld in the tank shell is required. Additionally, 10 percent of all tank welds must undergo ultrasonic, magnetic particle, or dye penetrant testing; and welds on reinforcement rings are subject to 100 percent ultrasonic, magnetic particle, or dye penetrant testing. Unless the design pressure exceeds 250 psi, production testing, including tensile, bend, and Charpy testing, for each 150 linear feet of tank seam weld must be conducted.

Although the aforementioned leak tightness test options are available, it is permissible to conduct this test in conjunction with the conventional hydrostatic strength test at 1.5 times the design pressure.

The nondestructive testing programs described above are intended to verify design assumptions by addressing the salient points of each tank design. The standards for conducting the various tests are identified in the proposed regulations and are based upon existing Coast Guard requirements or industrial consensus standards.

With a containment system which is well designed, constructed in accordance with approval procedures, and evaluated nondestructively in a manner commensurate with the tank type, the risk of primary barrier failure and possible resultant damage to the hull are greatly reduced. Protection of the ship's hull is of primary concern.

HULL STEEL

As indicated earlier, a primary objective in containment system design is the protection of the ship's hull from the cryogenic cargo. This objective is shared by those portions of the proposed standards which deal with the requirements for hull steel.

The hull steel section addresses design and construction standards for both the outer hull and the contiguous hull structure. For the most part the selection process for either application involves investigation of a specified design condition, determination of the resulting design temperature, and selection of the appropriate material from the standards of a recognized classification society. Domestically, the grading of steel to conform with this selection process is performed by the American Bureau of Shipping (ABS). (10) The ABS grading system, consistent with those of the other recognized classification societies, consists of five basic hull steel grades, increasing in extent of quality control from Grade A to Grade E. Grades D and E, which offer guaranteed notch toughness properties, play an important role in the Coast Guard requirements for outer hull steel.

Although great progress has been made toward complete international agreement on the IMCO Gas Code, there are certain areas where we still feel additional action must be taken to protect our ports and contiguous population. One of these areas is outer hull steel selection.

The rules of many classification societies, forming the basis for selection of most outer hull steels, today permit the construction of 165,000m^3 LNG ships with the entire outer shell made of Grade A steel. No strakes of material having enhanced notch toughness properties to act as conventional crack arrestors are required. The rules of most classification societies for crack arrestors are based on a correlation between material thickness, material toughness, and ship length. However, the gas ship introduces two new considerations to this question that are not covered by their rules. First, there is the ever present danger and history of nuisance spills due to flange or valve leakage or operating error. This will result in cracking of the deck, etc., but, in the presence of proper crack arrestors, the cracks will not propagate to a catastrophic failure of the hull girder. This danger is not a function of vessel length. Second, the double hull structure of gas ships, shown in Figure 3, leads to shell and main deck plating that is sufficiently thin to permit Grade A steel to be used throughout. The current rules of the societies were not based on experience with, and consideration of, double hull structures.

Resolution of this problem by the classification societies is being actively pursued and some intermediate remedies have been advanced. However, in the absence of a technically acceptable alternative approach to this problem, the U. S. Coast Guard has adopted the following requirements for crack arrestors for gas ships regardless of vessel size or type of containment system:

Sheer strake and deck stringer - Grade E or better

Bilge strake - Grade D or better

The above requirements remain applicable to new U. S. flag gas ships and to those of foreign flag entering our ports. Additionally, for those vessels carrying cargo which, under conditions of primary barrier failure and with ambient temperatures specified by a recognized classification society, causes the outer hull steel temperature to drop below 0°C, the material selection criteria normally applied to contiguous hull structure are applied to the outer hull as well.

The contiguous hull structure, shown in Figure 4, is that portion of ship's structure, including inner bottom plating, longitudinal and transverse bulkhead plating, floors, webs, stringers, and attached stiffeners, in closer proximity to the cargo containment system than the outer hull. The

contiguous hull structure, particularly that portion which
intervenes between cargo tanks, will experience, under
operating conditions, service temperatures considerably
colder than those experienced by the outer hull. Further,
since most containment system concepts necessarily include
a secondary barrier which is designed to contain cargo liquid
in the event of primary barrier failure, the situation described above is aggravated by the damaged or emergency condition. When considering ship survival, the contiguous hull
structure must be capable of performing under either of these
circumstances. To that end, a heat transfer calculation must
be submitted for the more severe (damaged) condition, the
solution of which results in either the selection of grade of
steel listed as acceptable for that service temperature in
the regulations or the installation of a special heating
system to maintain a service temperature which provides more
flexibility in material selection.

In arriving at the service temperatures for contiguous
hull structure, values for ambient air and sea temperatures
had to be assumed. These ambient design temperatures for
worldwide unrestricted service appear in the IMCO Gas Code,
but differ from those in the Coast Guard Proposed Standards.

The U. S. Coast Guard, based on a statistical analysis
of U. S. weather data, had consistently required the following
design ambients:

Lower 48 States

Air (at 5 knots):   -18°C
Sea Water: 0°C

Alaska

Air (at 5 knots):   -29°C
Sea Water: -2°C

Other nations felt that a +5°C still air ambient was
satisfactory for worldwide service. The resolution at IMCO
was to adopt the following values for design ambients, but
to permit the individual jurisdictions to use these or other
values, provided they were indicated on the vessel's IMCO
Certificate of Fitness.

IMCO Worldwide

Still Air: +5°C
Sea Water: 0°C

The U. S. Coast Guard was not prepared to accept these figures for unrestricted service, but was willing to reexamine its data in a much more comprehensive manner to see under what conditions +5°C would be acceptable. A statistical study examining the ambient air and sea temperatures in major U.S. shipping ports, was undertaken and resulted in a proposal to subdivide U. S. coastal areas according to similar ambient conditions. (11)   Implementation of this study would permit restricted (by month) service to the U. S. by those vessels designed using the IMCO ambients. For the present, however, except in the case of outer hull steel selection, the Coast Guard design ambients for unrestricted service remain those indicated formerly.

Regarding the welding and nondestructive testing of hull steels, the only deviations from the classification society standards normally applied involve (1) special low temperature weld procedure qualification for contiguous hull structure with a design temperature colder than −18°C, (2) production weld testing for each 50 meters of full penetration butt joint on contiguous hull structure with a design temperature colder than −34°C, and (3) 100% radiographic testing for certain butt welds in those outer hulls which form part of a secondary barrier.

CONCLUSION

The Coast Guard has actively pursued the development of both international and domestic standards for the safe transportation of liquefied gases for many years. The adoption of the IMCO Gas Code formalized the efforts of the world's leading experts in the field and was free from the reductions in safety standards normally expected in negotiation.

The Proposed Standards implement this document and make its recommendations mandates within our jurisdiction. Considering the breadth of areas addressed in these regulations, the cargo tanks and hull are but a part, albeit a critical one, of the total picture. But with due care being exercised in all facets of a fully coordinated, systematic approach to liquefied gas ship design and construction, the ultimate goal of safe and successful vessel operation will be much easier to achieve.

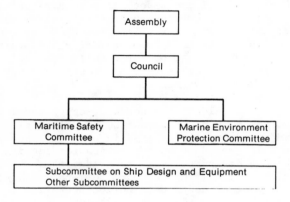

IMCO Committee Structure

(Abbreviated)

Figure 1

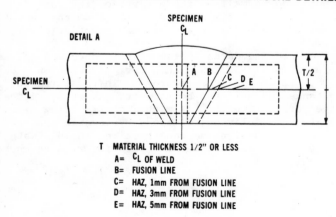

CHARPY V-NOTCH SPECIMEN REMOVAL DETAILS

T   MATERIAL THICKNESS 1/2" OR LESS
A= C_L OF WELD
B= FUSION LINE
C= HAZ, 1mm FROM FUSION LINE
D= HAZ, 3mm FROM FUSION LINE
E= HAZ, 5mm FROM FUSION LINE

Figure 2

LOCATION OF ENHANCED GRADES OF STEEL AS CRACK ARRESTERS

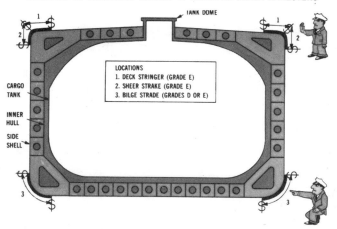

LOCATIONS
1. DECK STRINGER (GRADE E)
2. SHEER STRAKE (GRADE E)
3. BILGE STRADE (GRADES D OR E)

Figure 3

CONTIGUOUS HULL STRUCTURE

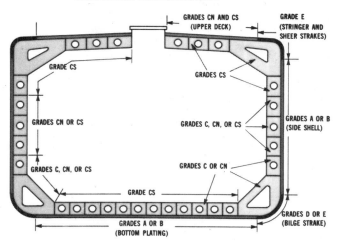

Figure 4

REFERENCES

(1) A. E. Henn and T. R. Dickey, "New Regulations for Liquefied Gas Carriers, "GASTECH 75 Proceedings, Paris, France, October 1975.

(2) J. F. Murphy and E. Quigley, "IMCO - A Forum for Maritime Nations," Surveyor, Vol. X, No. III, 1976.

(3) J. W. Kime, R. J. Lakey, and T. R. Dickey, "A Review of the IMCO Code for Gas Ships," SNAME Spring Meeting/ Star Symposium, San Francisco, CA, May 1977.

(4) "Proposed Standards for Self-Propelled Vessels Carrying Bulk Liquefied Gases," Federal Register, Vol. 41, No. 193, 1976.

(5) IMCO Resolution A.328(IX), "The Code for the Construction and Equipment of Ships Carrying Liquefied Gases in Bulk," 1975.

(6) "U. S. Coast Guard Marine Engineering Regulations," Title 46, Code of Federal Regulations, Parts 50-60, 1977.

(7) "Fracture Toughness Testing at Cryogenic Temperatures," ASTM Publication STP 496, 1971.

(8) "Fatigue and Fracture Toughness - Cryogenic Behavior," ASTM Publication STP 556, 1974.

(9) "Properties of Materials for Liquefied Natural Gas Tankage," ASTM Publication STP 579, 1975.

(10) "Rules for Building and Classing Steel Vessels," American Bureau of Shipping, 1977.

(11) J. G. Hicks and A. E. Henn, "Liquefied Gas Carriers - Statistical Analysis of Ambient Design Temperatures for the United States," GASTECH 76 Proceedings, New York, NY, October 1976.

TOUGHNESS CONSIDERATIONS FOR MERCHANT SHIP HULLS

Michael F. Wheatcroft
American Bureau of Shipping

ABSTRACT

This paper deals with brittle fracture prevention in merchant ships. Merchant ships are built to standards, known as Rules, established and administered by ship classification societies such as the American Bureau of Shipping (ABS). It is the primary function of a ship classification society to assure the soundness and seaworthiness of merchant vessels. The paper will review considerations given to design, material selection and application, welding effects, and inspection to forestall the occurrence of brittle fracture.

INTRODUCTION

Calamitous failures of engineering structures occur when the importance of some detail of design, material, fabrication, or operation goes unrecognized. Brittle fracture in ships, contrary to common belief, did not begin with the Liberty ships and T2 tankers built during World War II. Brittle fractures occurred in prewar riveted ships, but cracks in these ships were generally short because of the crack-arresting quality of the riveted joints; therefore little attention was given to the condition. With the advent of the welded ship, an essentially monolithic structure with multiaxial stress fields in the vicinity of welds was produced so that cracks potentially could travel farther. The importance of the change from riveting to welding went unrecognized and several Liberty ships and T2 tankers broke completely in two.

Investigation of World War II ship failures showed that

every fracture started at a notch (either a "design notch," such as a square hatch corner, or other cutout in a highly stressed area, or a weld flaw). It was also determined that, in many instances, the steel used was brittle at the service temperature, and that the highest frequency of fracturing occurred under a combination of heavy seas and winter temperatures.[1]

The frequency of brittle fracturing was reduced considerably during the latter part of the war by changes in structural design thereby reducing design notches (Fig. 1) and by improvements in weld quality thereby reducing metallurgical notches. Although it was recognized that notch toughness should also be improved, there was a lack of metallurgical knowledge as to what made structural steel sensitive to unstable fracture. Extensive studies of the metallurgical variables enabled ABS to lead the field in improvements in hull steel properties by 1947.

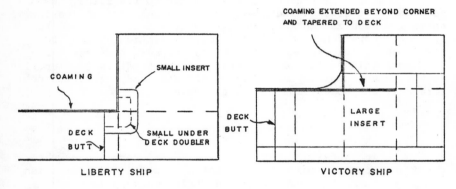

FIG.1 IMPROVEMENT IN HATCH CORNER DESIGN

Today, brittle fractures in merchant ships are rare or insignificant owing to careful attention to material selection and application,design details, fabrication techniques and inspection. Nevertheless, because of changes in marine technology and related fields,it is necessary to anticipate or respond promptly to conditions that may create a brittle fracture hazard. The development of liquefied natural gas (LNG) carriers is one case in point: Despite insulation of the cryogenic cargo, parts of the hull are cooled to such an extent that materials must be carefully selected, and special precautions taken in fabrication to retain adequate toughness of the weldments.

Another more visible example of change in the marine
field is the growth in tanker size. These supertankers are
among the world's largest structures. Some VLCCs (very large
crude carriers) carry enough crude oil on a single voyage to
make 40 million gallons of gasoline and are a quarter of a
mile long. As stresses in the hull girder increase with increasing ship length, necessitating use of thicker sections
or higher strength steels, careful consideration must be
given to toughness. ABS Rules, for example, provide for
material of high toughness in certain locations of all ships
over 300 feet in length to act as crack arresters and to reduce
crack initiation risk.

THE MERCHANT SHIP HULL

In addition to being the largest type of mobile structure
ever used in a transportation system, the modern merchant ship
is one of the few structures that relies principally on its
"skin" to provide the necessary strength for resisting operational loads. The skin, usually steel plates, welded and
stiffened in the approximate form of a box girder, must have
openings, usually in the deck, for operational reasons. These
openings may be very small, such as in oil tankers, or they
may be very large, occupying almost the whole deck, as is the
case with containerships. The details and location of openings
in the hull girder must be designed with appropriate consideration to their potential to cause brittle fracturing. Such
considerations are handled at the design review stage which
is the first point in vessel development where effective steps
can be taken by ABS to prevent or reduce subsequent failures.
It is an important responsibility of a classification society
to verify that submitted designs adhere to accepted standards
of good practice for vessel design as embodied in its Rules.
Through such a procedure ABS can identify design details which
may not conform to accepted standards and which may be a
threat to brittle fracture.

Consideration for toughness in a ship hull is essentially
confined to welded regions subject to tensile loading. In
general the areas of prime concern are the strength deck and
bottom shell of the ship in the midship 4/10ths portion, referred to as the 0.4L. This portion of the ship is subject
to maximum bending moments. When the ship is centered on a
wave crest, the deck is in tension and the bottom in compression; this is referred to as a "hogging" condition. When
the ship is centered on a wave trough, the reverse condition
prevails and the ship is said to be   "sagging" (Fig. 2).

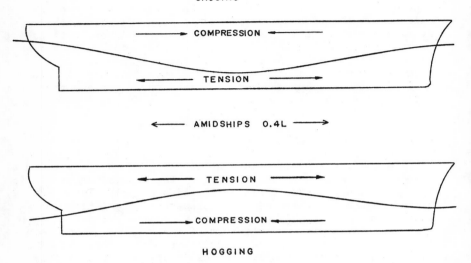

SAGGING

AMIDSHIPS 0.4L

HOGGING

FIG. 2 IDEALIZED WAVE ACTION

Tensile and compressive stresses in the hull vary from extreme levels in the deck and bottom to practically zero at the neutral axis. Tensile and compressive bending forces are greatest on the elements farthest from the neutral axis, usually the junctions of the deck and bottom with the side shell, known as the sheer and bilge strakes. Fig. 3 shows the distribution of "special material" (high toughness steel) in the mid body cross section of tankers and bulk carriers over 800 feet long, as required by ABS Rules. Unless mass and buoyancy are identical, a bending moment will act on the hull even without wave action: this is known as the still-water bending moment. Total bending moment is the sum of the still-water bending moment and the wave-induced bending moment. Gross response to bending moments will be deflection of the hull girder. Hull deflection is restricted by ABS Rules, which requires special consideration to ocean-going ships in which the length-to-depth ratio is greater than 15 to 1.

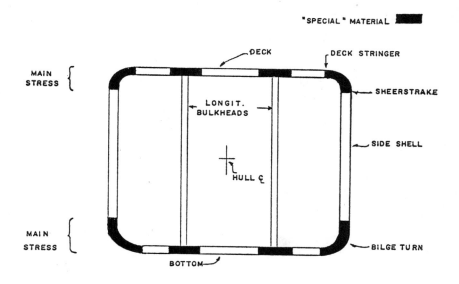

"SPECIAL" MATERIAL

DECK

DECK STRINGER

MAIN STRESS

SHEERSTRAKE

LONGIT. BULKHEADS

SIDE SHELL

HULL ₵

MAIN STRESS

BILGE TURN

BOTTOM

SIDE SHELL

FIG. 3 IDEALIZED MID-BODY CROSS-SECTION OF LARGE TANKER

Besides still-water and wave-induced bending loads there are other static and dynamic loads imposed on a ship, for example, shipped-water loads, and bow slamming. Additionally, oblique seas may impose torsional moments on the hull, while in head seas, steep short crested waves tend to induce hull girder vibrations. A ship is never in a no-load condition, nor even in a condition in which the absolute value of the longitudinal bending moment is known exactly[2].

It is a principle of naval architecture that as ship length (L) increases, the midship deck and bottom section modulus, and consequently in a general way plating thickness, must increase in proportion to L^2 to maintain the same stress level up to a length of 1000 feet. In ships over about 1000 feet long, however, the hull girder stresses increase at a somewhat lesser rate, and midship section modulus goes up only in proportion to L. This is because when ocean wavelengths become comparable to ship length, they do not achieve heights proportionally as great as those associated with shorter wavelengths[3]. This is advantageous from a fracture viewpoint because very thick plate would otherwise have to be used in long ships. This leveling off of the required midship plate thickness makes it possible to use hull steel for even the largest ships, in a thickness range where constraint is not a major influence.

It should perhaps be noted that the actual in-service loading on a ship hull is determined statistically: the seaway is constantly changing in an unpredictable manner, and other factors such as ship speed, heading, and weight distribution can have significant effect on hull girder response. Because of the variation of in-service loads the general rules for proportioning a ship hull section modulus are largely empirical. However, increased use of computers and analytical techniques is giving a better understanding of hull girder response; at the same time oceanographic research is increasing our understanding of ocean wave spectra. Given the above it is foreseeable that ship design will eventually be based more on scientific than empirical evidence. Empirically derived safety factors in the meantime provide a wide enough margin between demand and capability for merchant ships so that the risk of failure is essentially related to the risk of brittle fracture and not insufficient longitudinal strength.[4] Thus, it can be said that a ship fractures in a brittle manner or it does not fracture at all.

Merchant ship design, in fact virtually all engineering design, involves some compromises in terms of meeting design objectives, cost and service goals. Normally compromises between technical and economic considerations are foremost. This applies no less to the consideration for toughness. From fracture mechanics we know that the tougher the material the greater the number of combinations of stress and flaw size a structure can tolerate without fracturing. Toughness, however, costs money. It is almost a natural law that "better means more expensive." This is certainly true for toughness: The difference in cost between the least and most expensive ABS hull steel grades is about $150/ton. As 40,000 tons of structural steel may be used in a big ship, it is important to optimize performance requirements taking into account the economic aspect.

The risk of brittle fracture in today's merchant ships is very small because of the following reasons:

1. Material - Considerable increase in ship steel toughness in the past thirty years (Fig. 4)
2. Design - Awareness of the role of ship design and structural details in fracture sensitivity

3. Flaws - Improvements in weld quality, and use of
   sensitive flaw detection methods
4. Stresses - Better knowledge of actual in-service
   stresses

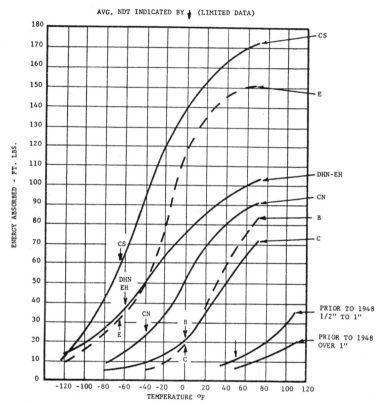

FIG. 4. AVERAGE CVN CURVES -ABS STEELS

Despite improvements, ship steel is still prone to fracture
in a brittle manner under certain conditions; this, however,
is not to be considered an indictment of its use. Its success
as a structural material is axiomatic. Brittle fractures
that do occur are inevitably the result of several adverse
conditions or mistakes; for example, a large undetected flaw
associated with a metallurgically degraded area, in a poorly
executed structural detail.

In general, ships are designed with both safe-life and

fail-safe considerations in mind; that is, although designed
to minimize the risk of fracture initiation, it is recognized
that flaws will exist and therefore may extend slowly by
fatigue, or rapidly in a brittle manner. All ship hulls are
subject to fatigue loading. There has, however, been little
evidence that fatigue is an important consideration in the
design of the primary hull girder structure. Further, fatigue
has not been identified as the cause of catastrophic cracking
of hull girders, although local fatigue fractures do occur in
ships. However, design for fatigue may become important with
increasing use of high-strength steels, because crack growth
rates go up and critical flaw sizes go down as yield strength
is increased.[5] The recognition of the existence of cracks
means that fail-safe concepts must be applied, by using tough
material in strategic locations.

In any structure built of materials, such as ship steels,
that exhibit a ductile-brittle transition over a temperature
range, service temperature must be considered. Except for
certain special situations the minimum service temperature
for ships is 0C (32F) (Fig. 5). In LNG carriers, however,
the cooling effect of the cargo contained in special insulated
tanks at -160C (-260F) on the hull must be considered. The
temperature of certain portions of the hull, may, depending on
the effectiveness of the insulation, reach as low as -40C
(-40F). Mobile offshore drilling units operating in arctic
climates, and having a significant part of their structure
high above the waterline, precluding any warming from the sea
or the unit, may also have minimum expected service temperatures as low as -40C (-40F). ABS has issued specific material selection guidelines for low-temperature mobile offshore
drilling unit operations.[6] Normal merchant ship operation,
(excluding LNG and refrigerated cargo vessels) however, is
considered to be at or above 0C (32F).

SHIP STEEL

The selection and application of appropriate materials
plays a crucial role in creating a successful ship. Mild
steel has been virtually the sole material for merchant ship
hulls since the early part of this century. The first steel
ship to cross the Atlantic was the "Banshee," built at
Liverpool in 1862 for use as a blockade runner in the American
Civil War. Early hull steel relied heavily on carbon as the
alloying element to achieve strength. Specifications required
only that the steel meet certain strength and ductility

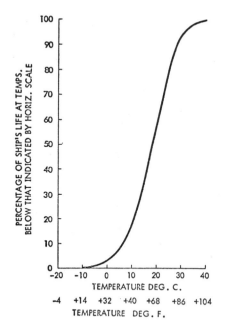

FIG. 5. Distribution of Service Temperature for Ships (Ref.20)

requirements; consideration did not have to be given to weldability or toughness.

Specification Development

The first published considerations for hull steel toughness were given in ABS's 1948 Rules. The Bureau's prescription
included controls on steel chemistry, deoxidation practice,
and thickness. The steels (Grades A, B, and C) specified in
the 1948 Rules also had application restrictions that were
determined by toughness considerations.

Chemical composition (steel chemistry) was controlled by
restricting the carbon content, specifying a certain C/Mn
ratio, and restricting impurities (sulfur and phosphorus).
Replacement of carbon with manganese had two beneficial
effects: acicular transformation products (brittle) were
bypassed, and liquation of weld fusion zone grains by sulfur

was prevented by the preferential formation of manganese sulfide. An important change was the requirement of fine-grain practice for Grade C. Although the full understanding of the effects of grain size on yield strength and toughness still awaited the work of Hall and Petch, fine-grain-size steel was known to be beneficial for toughness.

In 1953 a normalizing heat treatment was added to the prescription for Grade C for heavier thicknesses. ABS required this heat treatment in response to large increases in the size of ships, and thus thicker hull steel, which was more likely to fracture in a brittle manner because of increased constraint. The effect of normalizing is to refine the grain size and generally homogenize the microstructure; both are helpful in improving toughness.

There have been many changes in hull steel specifications since ABS first included toughness considerations in its 1948 Rules. These changes reflect developments in marine technology, developments in metallurgy, investigation of ship casualties, and efforts at unification between the world's major ship classification societies. The basic specifications for ABS grades of hull steel are shown in Tables 1 and 2.

TABLE 1

ORDINARY STRENGTH HULL STRUCTURAL STEEL

GRADES	A	B	D	E	CS	DS
PROCESS OF MANUFACTURE	FOR ALL GRADES: OPEN HEARTH, BASIC OXYGEN, OR ELECTRIC FURNACE					
DEOXIDATION	ANY METHOD EXCEPT RIMMED		SEMI-KILLED OR KILLED	KILLED, FINE GRAIN PRACTICE	KILLED, FINE GRAIN PRACTICE	KILLED, FINE GRAIN PRACTICE
CHEMICAL COMPOSITION (LADLE ANALYSIS)						
CARBON, %	0.23 MAX.	0.21 MAX.	0.21 MAX.	0.18 MAX.	0.16 MAX.	0.16 MAX.
MANGANESE, %	—— *	0.80-1.10	0.70-1.40	0.70-1.50	1.00-1.35	1.00-1.35
PHOSPHORUS, %	0.04 MAX.	0.04 MAX.	0.04 MAX.	0.04 MAX.	0.04 MAX.	0.04 MAX.
SULFUR, %	0.04 MAX.	0.04 MAX.	0.04 MAX.	0.04 MAX.	0.04 MAX.	0.04 MAX.
SILICON, %		0.35 MAX.	0.10-0.35	0.10-0.35	0.10-0.35	0.10-0.35
HEAT TREATMENT	——	——	NORMALIZED OVER 35.0 MM (1.375 IN.)	NORMALIZED	NORMALIZED	NORMALIZED OVER 35.0 MM (1.375 IN.)
TENSILE TEST						
TENSILE STRENGTH	FOR ALL GRADES: 41-50 KG/MM², 58,000-71,000 PSI					
YIELD POINT, MIN.	FOR ALL GRADES: 24 KG/MM², 34,000 PSI					
ELONGATION, MIN.	FOR ALL GRADES: 21% IN 200MM (8 IN.); 24% IN 50MM (2 IN.); 22% IN 5.65√A (A EQUALS AREA OF TEST SPECIMEN)					
IMPACT TEST STANDARD CHARPY V-NOTCH						
TEMPERATURE	——	——	−20 C (−4 F)	−40 C (−40 F)	——	——
ENERGY, MIN. AVG.	——	——	2.8 KGM (20 FT. LBS.)	2.8 KGM (20 FT. LBS.)	——	——
NO. OF SPECIMENS	——	——	3 FROM EACH 40 TONS	3 FROM EACH PLATE	——	——

* GRADE A PLATES OVER 12.5MM (0.50 IN.)
THE Mn SHALL BE 2.5 x C% (MIN.)

The higher strength steels (Table 2) were added to ABS Rules in 1966. These grades, which are representative of the "Microalloyed Steels," developed in the 1950's and 1960's, answered a demand in many areas of industry for tough, high-yield-strength steel with good weldability and cold formability at low cost. Their introduction to shipbuilding allowed reduced plate thickness (scantlings), which in turn allowed greater deadweight capacity and faster construction (shorter welding time and ability to handle larger sections). However, buckling

TABLE 2

HIGHER STRENGTH HULL STRUCTURAL STEEL

GRADES	AH 32 OR AH 36	DH 32 OR DH 36	EH 32 OR EH 36
PROCESS OF MANUFACTURE	FOR ALL GRADES: OPEN HEARTH, BASIC OXYGEN, OR ELECTRIC FURNACE		
DEOXIDATION	SEMI-KILLED OR KILLED	KILLED, FINE GRAIN PRACTICE	KILLED, FINE GRAIN PRACTICE
CHEMICAL COMPOSITION (LADLE ANALYSIS)	FOR ALL GRADES:		
CARBON %	0.18 MAX.		
MANGANESE %	0.90-1.60		
PHOSPHORUS %	0.04 MAX.		
SULFUR %	0.04 MAX.		
SILICON %	0.10-0.50 { AH TO 12.5 MM (0.50 IN.) MAY BE SEMI-KILLED		
NICKEL %	0.40 MAX. { IN WHICH CASE 0.10% MIN. Si DOES NOT APPLY		
CHROMIUM %	0.25 MAX.		
MOLYBDENUM %	0.08 MAX.		
COPPER %	0.35 MAX.		
ALUMINUM % (ACID SOLUBLE)	0.06 MAX.		
COLUMBIUM % (NIOBIUM)	0.05 MAX.		
VANADIUM	0.10 MAX.		
HEAT TREATMENT	NORMALIZING REQ'D. OVER 12.5 MM (0.50 IN.) IF Nb TREATED	NORMALIZING REQ'D. OVER 25.5 MM (1.0 IN.) IF Al TREATED OVER 12.5 MM (0.50 IN.) IF Nb TREATED OVER 19.0 MM (0.75 IN.) IF V TREATED	NORMALIZED
TENSILE TEST TENSILE STRENGTH	FOR 32 GRADE: 48-60 KG/MM^2 (68,000-85,000 PSI) FOR 36 GRADE: 50-63 KG/MM^2 (71,000-90,000 PSI)		
YIELD POINT, MIN.	FOR 32 GRADE: 32 KG/MM^2 (45,500 PSI) FOR 36 GRADE: 36 KG/MM (51,000 PSI)		
ELONGATION, MIN.	FOR ALL GRADES: 19% IN 200 MM (8 IN.); 22% IN 50 MM (2 IN.) 20% IN 5.65√A (A EQUALS AREA OF TEST SPECIMEN)		
IMPACT TEST STANDARD CHARPY V-NOTCH			
TEMPERATURE	———	-20C (-4F)	-40C (-40F)
ENERGY, MIN. AVG.	———	3.5 KGM (25 FT. LBS.)	3.5 KGM (25 FT. LBS.)
NO. OF SPECIMENS	———	3 FROM EACH 40 TONS	3 FROM EACH PLATE

has to be considered in areas of the ship hull that will experience compression loads; thus, full utilization cannot be made of increased strength because ability to resist compression loads is largely a function of geometry.

Special precautions must be taken when welding higher-strength steels, because of possible metallurgical degradation resulting in toughness loss in the weld heat affected zone (HAZ). This is especially true of the 70-kg/mm^2 (100-ksi) Quenched and Tempered steels with tempered martensite microstructures. Although not used extensively in ship hulls, they have been used for such applications as the box-girder section of containerships. In these ships hatch openings extend almost completely across the full deck width. Therefore to obtain adequate longitudinal strength in the narrow outboard deck plating, high-strength Q&T steels are an attractive consideration.

Anistropy

The toughness properties of wrought steels exhibit anistropy; that is, the properties depend on which direction they're measured relative to the rolling direction. Toughness is highest in the longitudinal direction, lower in the transverse, and lowest through the thickness (Fig. 6).

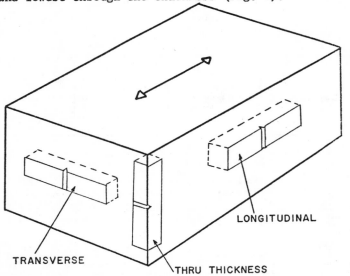

LONGITUDINAL

TRANSVERSE

THRU THICKNESS

FIG. 6. - ORIENTATION OF TEST SPECIMENS RELATIVE TO
PLATE ROLLING DIRECTION

The reason for the anistropy of properties is largely because of the elongation of nonmetallic inclusions, especially sulfides, in the direction of rolling. Generally, for ship steel, this has not been a problem because plates are oriented in the ship with the rolling direction normal to the primary loading axis (Fig. 7). As ships become wider in the beam, however, transverse stresses can be significant, and transverse toughness may have to be considered.

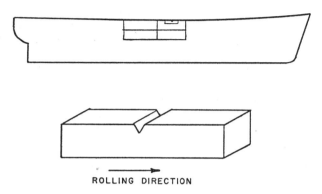

ROLLING DIRECTION

FIG. 7. - ORIENTATION OF SHIP PLATES AND CHARPY SPECIMEN
SHOWING ROLLING DIRECTION

Also, through-thickness stresses and strains from welding can result in steel separating along nonmetallic planes, a condition called lamellar tearing. This form of tearing, although it has not resulted in brittle cracks in ships, has been a considerable nuisance in tubular connections in mobile offshore drilling units. Sulfides, can also affect the risk of hydrogen induced cracking in heat affected zones of welds by acting as hydrogen "traps." It is now possible for steel mills to control sulfide inclusion morphology by the use of a rare earth addition, so as to give an essentially isotropic steel. Oxide inclusions on the other hand are less easily controlled. Nonmetallic particles, which are brittle compared with the metallic matrix, impair the metals' capacity to deform. This effect increases with increasing yield strength. Unfortunately nonmetallic inclusions are permanent features of steel. They vary only in amount and distribution. Their character depends on the steelmaking operation, the chemistry of the steel, and the type of deoxidizers used.

To reduce inclusions to a minimum, sulfur and oxygen contents of the steel must be reduced. Low levels of sulfur

can be specified from the steelmaker; low oxygen contents, by
specifying a fully killed (fully deoxidized, or vacuum-degassed
steel.) Inclusion shape control and vacuum degassing are
expensive, and justifications for specifying these measures
should be based on the particular application. For example,
the Bureau may recommend that consideration be given to through-
thickness properties if there is reason to believe that welding
strains in that direction could be significant.

Toughness Tests

ABS steel specifications have included Charpy V-notch
(CVN) tests since the early 1950's, but still retain prescription grades - grades with prescribed processing and
chemistry to ensure toughness but with no toughness test
required. The validity of the prescription approach is based
on the premise that control of the manufacturing process,
deoxidation, heat treatment, chemistry, and so forth applies
to the whole batch of steel produced, whereas a toughness test
measures a tiny area, and thus may not be representative of the
whole batch from which it is taken. For example, Grade CS,
a prescription grade, and Grade E, a CVN tested grade, are
both acceptable for identical applications, and, as Fig. 8
shows, the toughness properties are very similar. Grade CS
prescription has a more favorable chemistry than E, which is
the reason for the slightly higher average toughness.

The CVN test is essentially an empirical test based on
experience and cannot be related directly to design. The
test has a number of drawbacks, not least of which is the
fact that its configuration absorbs a large amount of energy
in fracture initiation, whereas the most important property
for hull steel is resistance to crack propagation, as cracks
will initiate in welds but usually propagate in the base metal.
The advantages of the CVN test, that it is simple to make and
conduct and is universally recognized and understood, have so
far outweighed its disadvantages however.

There is disagreement among experts as to which test
would best supplant the CVN. More than fifty types or variants
of tests can be identified as having been used for evaluating
the susceptibility of material to brittle fracture. In response to the need to characterize crack arrest and propagation
in terms that can be directly related to design, a variety of
fracture mechanics tests have been applied to ship research.
Such tests are particularly useful for evaluating new materials

or new applications where correlative data of service performance with conventional impact tests is unavailable. There is,
however, no test that can guarantee absolute safety, especially
in such a complex welded structure as a ship. In the final
analysis, the service performance provides the desired
evaluation.

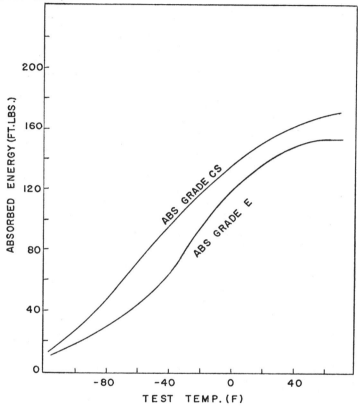

FIG. 8 AVERAGE CHARPY CURVES FOR ABS E & CS HULL STEELS

     The tests favored at the moment for evaluating ship steels
are the crack opening displacement (COD) test, generally
favored in Europe, and the dynamic tear (DT) test, which was
developed by the U. S. Navy.

     COD can be considered as the amount of inelastic stretching
of the material immediately ahead of the crack tip. The test
is superficially simple, although at present no accepted

operational definition exists to provide a basis for homogeneity among sets of data represented by various investigators.[8] The DT test (MIL STD 1601 (Ships)) began replacing the CVN test in U.S. Navy specifications for ship hull materials in 1970. This test measures the total energy to fracture a specified cross-section of limit severity test conditions and determines the characteristic fracture progagation mode.

One test that has been largely overlooked is the charpy pressed notch (CPN), which overcomes the disadvantage of the blunt-notched CVN and yet retains ease of sample preparation and testing. This is extremely important where high volumes of relatively inexpensive materials are involved - a consideration that tends to be overlooked by proponents of exotic fracture mechanic tests. Work at ABS with CPN has shown it to be a very reliable indicator of the nil-ductility transition (NDT) temperature of ship steels. The standard test method for determination of NDT temperature is described in ASTM E208. NDT temperature is the maximum temperature at which a steel fractures in a brittle manner under certain test conditions. The NDT temperature signifies a degree of brittleness such that small cracks are critical for fracture initiation.[9] Experience with service failures has shown that a service temperature $\geq$ 17C (30F) above NDT temperature provides reasonable assurance of crack arrest in structural carbon steels for stresses up to about half of the yield strength, and $\geq$ 33C (60F) above NDT temperature for stresses up to about yield point. Typical NDT temperatures for ABS steels range from -10C (+14F) for Grade B to -40C (-40F) for Grades E and CS. Measurements on large tankers and bulk carriers have indicated the maximum wave bending excursion (peak to trough) encountered to be about 10-kg/mm^2 (14,000 psi), or about one half the yield stress of ordinary strength ship steel.[10] At points of stress concentration the local stresses may reach yield point magnitude.

EFFECTS OF WELDING ON TOUGHNESS

Freedom from brittle fractures in a ship hull depends essentially on the toughness and integrity of the weldments. A weldment consists of weld metal, heat affected zone, and base metal.

Progress in welding techniques has been substantial, however application of new welding processes in shipbuilding has been relatively slow. A major reason for the slow development of automation in shipyards is that about three-quarters of the welds in a ship hull are fillet welds, which are often interrupted by intercostal structural members. Although large ships are now fabricated with considerable amounts of automatic and semi-automatic welding processes, manual shielded metal arc (SMA) welding still dominates the small yards and small ships. The high deposition rate electro-gas and electroslag (EG) (ES) processes generally give sounder welds than SMA but have a greater potential for metallurgical damage because of their high heat input.

In general welding may adversely affect toughness in two ways: metallurgical damage by undesirable microstructural transformation, and creation of stress concentrations by physical defects. These general effects may be aggravated by residual stresses induced by thermal conditions. Of the two possible detrimental effects of welding, the most danger-ous from a brittle fracture standpoint is metallurgical damage, at least in such inherently plastic materials as ship steels, which are quite tolerant of relatively large flaws.

The microstructure in the region of the flaw tip is one of the primary factors which determines whether a flaw will propagate in a brittle manner. Consideration must therefore be given to the microstructural changes induced by a particular welding process or procedure and the resulting effect on mechanical properties. The Bureau requires that procedures for the welding of all joints, that is, for the welding pro-cesses, types of filler metals, edge preparations, welding techniques, and positions proposed be established in writing by the shipbuilder. It is the Surveyor's responsibility to ensure compliance with the established procedures. To assure adequate toughness in weld metals, the Bureau has classed filler metals into six distinct grades and each grade is required to meet a toughness criterion comparable with a given base plate grade.

Thermal Effects

Metallurgically, welding is a heat treatment in which a particular base metal zone experiences very fast heating, and cooling from a peak temperature. In practice, this heat affected zone (HAZ) may show considerable loss of toughness if significant grain coarsening occurs. Significant grain coarsening is usually a problem only when single-run high-heat-input welding processes are used with heat-treated steels. Multirun procedures are usually beneficial in improving weld metal and HAZ toughness because of the grain-refining effect

of each weld pass on the previous deposit. Not only the degree
of metallurgical degradation but the extent is affected by
heat input in that the width of the HAZ is directly proportional
to the net heat input. Heat input is the quantity of heat
energy introduced per unit length of weld   from a traveling
heat source such as an arc. Heat input = power/travel velocity
of source. The high-heat-input welding processes used in shipbuilding are electroslag (ES) and electrogas (EG). Because
of the metallurgical degradation associated with EG and ES
welding, the Bureau requires shipyards to perform qualification
toughness tests to show that the HAZ can meet similar requirements to those of the base metal in high stress regions of the
ship.

As deposition rate is a direct function of heat input,
high heat input processes are attractive to shipbuilders who
want to deposit the minimum amount of weld metal at the fastest
possible rate. The builder may be restrained in the quest for
higher production rates by the toughness considerations of
regulatory bodies such as ABS. In the case of LNG carriers
EG and ES processes have not been qualified by ABS for welding
sections of the inner hull cooled by the cargo because of low
toughness levels in weld HAZ's exposed to low temperature.

The effort to solve the problem is being directed toward
improving base metal tolerance to high heat input. Fig. 9
illustrates how different welding processes affect toughness
of a microalloyed steel. The toughness values are taken from
a study by ABS for the Maritime Administration.[11]

Finally, EG and ES welding affect the toughness of the
weldments indirectly. The stress condition around the weld
is such that it will tend to retain a moving crack in the HAZ.
Also as the HAZ of high heat input welds is more extensive,
there is a statistically greater chance of cracks existing in
these HAZ's. With low energy multipass welds the stress field
is such that a moving crack in the HAZ will tend to propagate
into the generally tougher base metal.[12] This effect is shown
schematically in Fig. 10. Because of the above mentioned
toughness problems with EG and ES processes, they are confined
for the most part, to vertical side shell welds between the
sheer and bilge strakes, that is, in the relatively low stress
regions of the hull, constructed of as-rolled ordinary strength
steels, which are relatively insensitive to heat input.

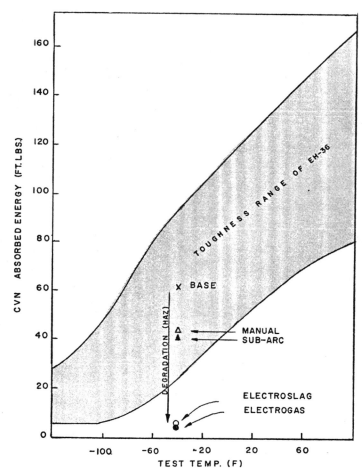

FIG. 9. EFFECT OF WELDING PROCESS ON TOUGHNESS OF ABS EH-36

Procedure Effects

Improper welding procedure or technique may result in a
number of types of flaws. Flaw here means a discontinuity
that may affect the stress field but not the metallurgical
structure. The principal flaws are planar and nonplanar.
Planar flaws, which include cracks, incomplete penetration,
lack of fusion, linear slag, and any other sharply notched
flaw. are more detrimental than nonplanar flaws, which include
porosity and dispersed slag.

How damaging a flaw can be depends on location/orientation, acuity, and size. In ship hulls, the location of flaws is of greater importance than size. A small flaw at a weld inter-section may be more detrimental than a large crack elsewhere. Planar flaws are especially detrimental when oriented perpen-dicular to the uniaxial loading axis; they have little effect when lying parallel to the axis. Flaw size (length) is of course important. From fracture mechanics we know that frac-ture toughness (K_I) is proportional to the square root of crack length. However, metallurgical influences from welding and flamecutting largely overshadow the influence of geometric factors such as crack length.

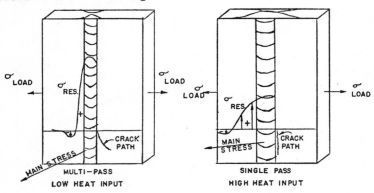

FIG.10. WELD STRESSES - EFFECT ON
CRACK PROPAGATION

Planar Flaws

Hydrogen-Induced Cracking. The sensitivity of steel to hydrogen embrittlement from welding increases with increasing yield strength; ordinary-strength hull steels are relatively immune to this problem. Atmospheric moisture and components in the electrode flux coating are the sources of hydrogen. Hydrogen cracking is often called delayed cracking because of the tendency for these cracks to initiate after the weld has cooled, sometimes several days after welding. For this reason it is common practice to perform final inspection several days after welding. Welds in which radiographs reveal any type of crack are considered unacceptable.[13]

Welding of high-strength steels requires the use of low-hydrogen electrodes, which, if not packaged in hermetically sealed containers, must be kept in ovens prior to use to control moisture content. As all coated electrodes are hygroscopic to some degree, they must be consumed within a short period of time after removal from containers or ovens. Preheating and maintaining a minimum interpass temperature also help to prevent the formation of weld metal and HAZ microstructures that are susceptible to hydrogen damage. Increasing the heat input also helps alleviate the condition, although this could lead to other problems, as mentioned previously under Thermal Effects.

Hot Cracking. Hot cracking is associated with thermal shrinkage stresses and tends to be prevelant in long continuous root passes where the weld area cannot sustain high stresses and thus cracks upon cooling. This type of cracking can be minimized by increasing the size of the root weld. Hot cracking is also sometimes seen in electroslag welds, and in single-pass fillet welds where cracks occur through the weld throat.

Incomplete Fusion. Incomplete fusion, which is the failure of the weld metal to fuse to the base metal or adjacent layers of weld metal, may occur in any part of the joint. Incomplete fusion may be caused by poor fit-up, low welding current, or improper welding technique.

Slag. Small, well dispersed slag inclusions are generally not harmful, but if slag is entrapped between weld metal and base metal or otherwise results in long or thin, sharp unfused areas, it may be harmful. Insufficient cleaning of weld passes, improper welding technique, and low current are the common causes of excessive slag entrappment.

Nonplanar Flaws

Porosity. Voids from gases released by the cooling weld metal or from chemical reaction in the weld are usually spherical and generally harmless; if dispersed. Porosity is affected by moisture, welding current, speed, and arc length, and can occur where the arc is struck especially when using low-hydrogen electrodes which require a short arc. (See also Slag.)

Repair Welding

Judgment must be exercised in deciding which flaws to

repair as the repair of harmless three-dimensional flaws may introduce more harmful, and less easily detectable planar flaws. ABS Rules require that all welded repaired areas be inspected as required by the Surveyor to determine acceptability of the repair weld.[13]

Although many flaws are in themselves harmless, if the structure contains many of them, there is apparently something wrong with the procedure or technique and an investigation would be in order. It is important to realize that as nondestructive testing becomes widespread, and sensitivity of the techniques improved, more flaws will be detected. In such a case the exercise of proper judgment will be particularly important in avoiding unnecessary repairs.

INSPECTION

Vessels classed with the American Bureau of Shipping are built under the supervision of Bureau Surveyors. The importance of surveillance during construction cannot be overemphasized. Ships have been lost because of poor construction.[14] It is an important function of ABS, through its field Surveyors, to maintain standards of quality.

In terms of preventing brittle fractures probably the most important aspect of the Hull Survey is to ensure the quality of the welds where cracks tend to originate. Surveillance of joint preparation, fit-up, and welding and nondestructive inspection procedures are important tasks of the Surveyor.

Visual inspection is generally satisfactory for fillet welds. During the welding operation itself visual inspection is important to ensure that proper slag removal and cleaning is carried out between passes. Also, as most joints of hull plates are welded from both sides, visual inspection of the joint after back-gouging of the first side would reveal any unfused areas, slag inclusion, or other root defects that remain because of improper gouging procedure.[15] Magnetic particle and dye penetrant inspection are used effectively in this instance.

The most dangerous flaws are those that are most difficult to detect. For instance, a hairline crack in a high-stress region of the hull, such as the deck, is far more hazardous than a gross slag pocket in a side shell weld.

The different methods of nondestructive testing have differing sensitivities to different flaw geometries. For example, radiography is more sensitive to nonplanar flaws such as porosity than to fine cracks, whereas the reverse tends to be true for ultrasonic inspection.

The Bureau's Rules for nondestructive inspection of hull welds express the extent of inspection as a function of ship size and location of welds within the structure. The number of checkpoints for radiographic or ultrasonic inspection within the midships 0.6L are expressed by the formula:

$$n = L \frac{(B+D)}{46.5}$$ where L, B & D are ship length, breadth, and depth expressed in meters, or

$$n = L \frac{(B+D)}{500}$$ where expressions are in feet

From the formula it is evident that the entire weld length is not inspected and thus it must be assumed that flaws may exist in the completed vessel.

In some specialized sections of a ship, the amount of inspection is increased. For example, all full-penetration welds in primary containers for cryogenic cargoes @ -73C (-100F) or colder are to be 100% inspected. For cargoes above -73C (-100F), 100% of the intersection of butts and seams are to be inspected.

Provision is made to define locations where inspection is to be concentrated. Within the midship 0.6L inspection is to be concentrated in locations of high stress and high-stress concentration as follows:

1. Intersections of welds perpendicular to the ship's axis (butts) with welds parallel to the ship's axis (seams) in the following areas:

    a) Sheer strakes (strength-deck side shell)
    b) Bilge strakes (curved bottom of side shell)
    c) Deck stringers (outboard-deck plating)
    d) Keel plates (bottom centerline plates)

2. Deck butts in vicinity of hatch corners

3. Butts in vicinity of breaks in the superstructure. Outside the midship 0.6L, inspection is to be performed at random according to the Surveyor's discretion. Inspection is

concentrated on butt welds because the principal loading axis
on a ship hull is perpendicular to those welds.

For both ultrasonic and radiographic inspection two
classes of criteria are used for acceptance of flaws:

1. Class A is intended for critical locations in the
0.6L of ships 150 m (500 ft) or larger, or in small ships
if justified by design or material. Class A criteria are
also specified for full-penetration welds in way of cargo
tanks of LNG and LPG carriers.

2. Class B criteria are intended for less critical
locations and for applications where Class A would not be
required.

RELIABILITY OF MERCHANT SHIPS

As with almost every worthy human endeavor, ship transportation involves risks. Considering the sometimes hostile
environment in which a ship must operate (approximately 50%
of the world's ship tonnage is concentrated in the stormy
North Atlantic), losses are commendably low. Tonnage losses
for all causes have remained relatively constant over the
past several years, averaging about 0.35% of total tonnage
at risk. News of spectacular shipwrecks notwithstanding,
the ships most at risk are small, old, coastal traders.[16]
This is in large part because of the danger of stranding and
the short steep seas encountered on continental shelves, that
are more likely to overwhelm the small ship.

Despite improvements in weather forecasting that may
enable a ship's captain to avoid some of the worst storms,
all ships, especially those plying the North Atlantic and
the Southern Oceans, will encounter heavy weather at some
time. The more rigorous the sea the greater the stresses
on the hull. The total wave force is a result of wave motion
plus ship speed. Stresses can be reduced as much as 50% by
speed reduction or by a change in heading, that is, by the
exercise of good seamanship.[17] It is understandable, however,
that ship officers will keep their ships at optimum speed as
long as possible. With today's enormous ships the captain
cannot "feel" the ship motions and therefore may not be fully
cognizant of wave loads on the hull.

ABS has carried out force measuring experiments on a
number of ships in service. The data generated is used to

confirm or modify existing Rules. In the future, large ships
may be equipped with hull surveillance systems to monitor
actual stresses in critical areas of the vessel. Limiting
wave bending loads may also be specified, in much the same
way as wing loads on aircraft are limited by restricting
performance.[2]

Are ships safe enough? How safe is safe enough? Generally
speaking society will tolerate certain improbable risks in order
to reap the benefits. The risk, for example, associated with
transporting LNG to Staten Island has been estimated to be
about one fatality every 10 million years for people living
and working along the approach route to the harbor.[18] That
level of risk is about the same as the risk of being struck
by lightning. And yet, there is a great opposition to off-
loading facilities for LNG ships. This is in part because,
although statistically the chances of an accident may be low,
the consequences of a single accident could be catastrophic.
Major ship casualties because of structural failures do not
occur frequently. By making various assumptions, Dalzell[19]
estimates that, the probability of structural failure is much
less than .003 for any given ocean-going ship over its life.
He then suggests a target figure for the risk of catastrophic
structural failure of .001 or one chance in a thousand that
a ship will be lost during a normal lifespan.

Although there are few recently published statistics
on brittle fractures in ships, an in-house study by ABS,
covering 26,880 ship-years (ocean-going ships in service
between 1967-1975), showed a total of 104 cracks in the 0.4L
sheerstrake, side, and bottom of general cargo, bulk and oil
carriers. None of these cracks were vessel endangering, and
they were not necessarily all brittle cracks.

Although it might seem that the reliability of a structure would improve in proportion to increased factors of
safety or safety margin, this is not so. Assume that we know
three things: (1) the range of external conditions that a
particular component "demands," (2) that the frequency distribution is normal (Fig. 11 bottom), and (3) that the toughness of the component, as determined from tests, is also
normally distributed (the "capability" curve in Fig. 11).
When the two curves are plotted, as in Fig. 11, a portion of
the areas under the curves overlap, as shaded, indicating the
possibility of failure. The area under the capability curve
is the reliability. The ratio of the mean T to the mean t
can be defined as a factor of safety. When this ratio increases, the shaded area will decrease and the reliability

will improve. But if we plot the difference between the demand and capability means (T-t)against reliability, we would obtain a curve shaped approximately as shown at the top of Figure 11. The curve shows that where reliability is already high, incremental increases of ΔT become less effective and more costly in improving reliability as the curve approaches the limit asymptotically.[21] Although these curves are only illustrative, they are presented to show that arguments for raising toughness standards in order to increase safety could lead to considerable increases in cost with little improvement in reliability.

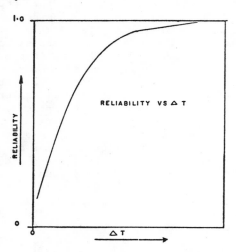

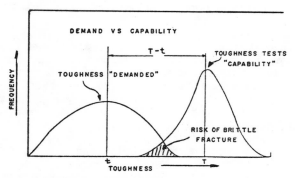

FIG. II.   DEMAND, CAPABILITY, & RELIABITY

SUMMARY

Careful attention to design, fabrication, and materials
has reduced to a low incidence the occurrence of brittle fracture in merchant ships hulls. This, however, is no time for
complacency; it is a time of rapid technological change, and
the potential for trouble is high when something new is
attempted. The danger is that the extent of the novelty may
well escape attention. The importance of close cooperation
between the designer and the materials and welding specialist
cannot be overemphasized.

ACKNOWLEDGMENTS

The author is grateful for the advice of his colleagues
in the preparation of this paper, particularly to R. Curry
for useful discussions.

REFERENCES

(1) Final Report of A Board of Investigation "The Design and
    Methods of Construction of Welded Steel Merchant Vessels,"
    Government Printing Office 1947.
(2) E. V. Lewis, et al, "Load Criteria for Ship Structural
    Design," Ship Structure Committee Report SSC 240, 1973.
(3) H. Shinto, "Big Ships," Science and Technology, March 1968.
(4) J. J. W. Nibbering, "Permissible Stresses and Their
    Limitations," Ship Structure Committee Report No. SSC 206,
    1970.
(5) P. Francis, J. Lankford, and F. Lyle, "A Study of Sub-
    critical Crack Growth in Ship Steels," Ship Structure
    Committee Report No. SSC 251, 1975.
(6) B. Alia, M. Wheatcroft and C. Null, "Material Selection,
    Welding, and Nondestructive Testing Guidelines for Offshore
    Mobile Drilling Units," Presented at 1976 Offshore Technology
    Conference, Houston.
(7) P. Hart, R. Dolby, N. Bailey and D. Widgery, "The Weld-
    ability of Microalloyed Steels," Microalloying Symposium,
    Washington, D. C., October 1975.
(8) "Rapid Inexpensive Tests for Determining Fracture Toughness,"
    National Academy of Sciences, Washington D. C., 1976.
(9) W. Pellini, "Principles of Fracture-Safe Design," Part 1,
    The Welding Journal, March 1971.
(10)S. T. Rolfe, D. M. Rhea, and B. O. Kuzmanovic, "Fracture
    Control Guidelines for Welded Steel Ship Hulls," Ship
    Structure Committee Report No. SSC 244, 1974.

(11) B. Alia, I. L. Stern and C. Null, "Toughness Evaluation of Electrogas and Electroslag Weldments," The National Shipbuilding Research Program, U. S. Maritime Administration, March 1975.

(12) J. J. W. Nibbering, "Fracture Mechanics and Fracture Control for Ships," Report No. 178S, State University of Ghent, May 1973.

(13) American Bureau of Shipping, "Rules for Nondestructive Inspection of Hull Welds," 1975.

(14) J. H. Evans, Editor "Ship Structural Design Concepts," Chapter 13, "Margin of Safety," (Caldwell) Dept. of the Navy Contract No. N00024-71-C-5173, June 1974

(15) American Welding Society, "Guide for Steel Hull Welding," AWS D3.5-76, February 1975.

(16) W. J. Beer, "Analysis of World Merchant Ship Losses," read at Royal Institution of Naval Architects in London, March 1968.

(17) J. Mentzoni, "The Ship and the Invincible Ocean," Veritas No. 76, Volume 19, October 1973.

(18) E. Drake, and R. Reid, "The Importation of Liquefied Natural Gas," Scientific American Volume 236, No. 4, April 1977.

(19) J. Dalzell, "Some Conjectures as to Acceptable Levels of the Risk of Structural Failure," (Unpublished Memo to members of U.S.A. Ship Research Committee, 12 May 1970).

(20) J. Hodgson, and G. Boyd, "Brittle Fracture in Welded Ships - An Empirical Approach from Recent Experiences," Quarterly Transactions, The Royal Institution of Naval Architects, Volume 100, No. 3, July 1958.

(21) A. F. Madayag, Editor, "Metal Fatigue: Theory and Design," Pg. 157-161, John Wiley and Sons Inc., 1969.

ACOUSTIC EMISSION TO DETECT FATIGUE
DAMAGE IN JET ENGINE MATERIALS

J. Stephen Cargill, Raymond M. Wallace
and Charles T. Torrey

United Technologies Corp., Pratt & Whitney Aircraft Group,
Government Products Division, West Palm Beach, Florida 33402

SUMMARY

Acoustic emission multiparameter distributions, including signal pulse width, peak amplitude and ringdown counts per event, were
conducted during elevated temperature strain control low-cycle
fatigue testing of IN-100 to characterize the various stages of
fatigue damage in the high-strength powder-metallurgy nickel-base
superalloy.

INTRODUCTION

Acoustic Emission (AE) monitoring has been established as a
viable method for detecting fatigue crack activity in high-strength
gas turbine disk alloys (1). Supplementary to the classical ringdown counts signal analysis methods, research is now underway to
determine additional information in each AE signal through frequency
domain and time domain analyses. This investigation is based on the
time domain approach, including AE signal pulse width, peak amplitude and ringdown counts (per event) analyses, to detect and classify
fatigue damage occurring during elevated temperature, strain control
Low-Cycle Fatigue (LCF) testing of IN-100.

Although the basis of this investigation is LCF strain control
specimen testing, results are expected to be applicable to complex
turbine engine components of the same material, whose fracture cri-

(1) Wallace, R. M., "Investigation of Fatigue Crack Propagation
in Ti 6Al-2Sn-4Zr-6Mo and IN-100 Using Acoustic Emission
Techniques," Pratt & Whitney Aircraft GP 74-236, Dec. 1974.

tical regions experience similar strain cycling. Future testing
and analysis will include laboratory cyclic loading of IN-100
engine components while performing acoustic emission Multiparameter
Distribution (MPD) analyses. The program result will be an experimental method to determine the extent of fatigue damage incurred in
a gas turbine disk component during its operational engine lifetime
and a methodology to predict useful residual life.

EXPERIMENTAL METHODS

Laboratory Apparatus

LCF strain control testing was performed using the specimen
configuration illustrated in Figure 1 by a servocontrolled, hydrauli-
test machine developed at Pratt & Whitney Aircraft Group's Governmer
Products Division. Particular attention was focused upon isolating
the test load frame from extraneous vibration originating in the sy-
tem hydraulics, etc.

All AE laboratory apparatus used in the development of multiparameter distribution analyses was supplied by Dunegan/Endevco of
San Juan Capistrano, California. The AE monitoring system and its
configuration are illustrated schematically in Figure 2. The capability to simultaneously monitor an AE event with three separate di-
tribution analyzers was initially developed for Pratt & Whitney Air-
by Dunegan/Endevco.

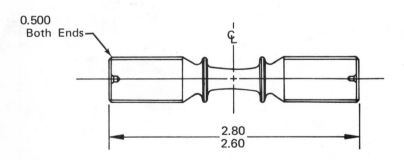

Figure 1. CONSTANT STRAIN LOW-CYCLE FATIGUE SPECIMEN

FIGURE 2

ACOUSTIC EMISSION MONITORING CONFIGURATION FOR
LCF STRAIN CONTROL TESTING

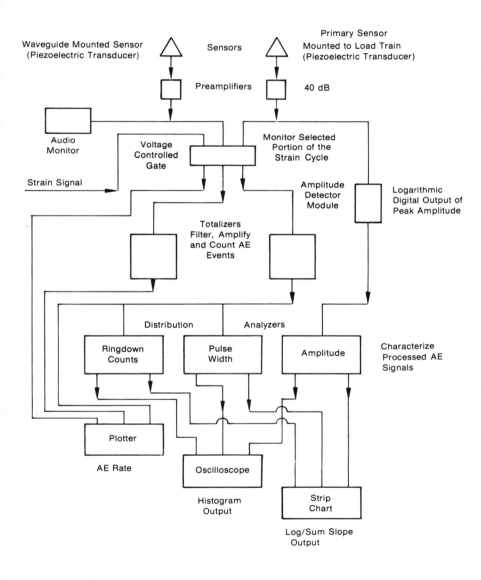

Test and Monitoring Procedures

LCF strain control testing of IN-100 was conducted at temperatures and strain levels which simulated localized operating conditions at the fracture critical location in an advanced gas turbine disk.

During initial AE monitoring, adjustments were made to the sampling parameters, including system gain, amplitude threshold and voltage controlled gate (operating from the strain signal) to establish the optimum balance between system sensitivity and rejection of extraneous mechanical and electrical noise. Adjustments were also made to distribution analyzer processing parameters (multiplier and signal envelope) to establish operating ranges which best characterize the AE signals associated with IN-100 fatigue cracks.

Two sensors were used throughout this test program to assess relative sensitivities of various waveguide configurations (Figure 3 Figure 4 illustrates waveguide attachment (through the upper load rod) and placement of the transducers. The lower sensor, mounted directly to the lower load rod, was the primary sensor from which all acoustic emission time domain distributions were performed. It was also used as a sensitivity reference upon which relative waveguide effectiveness was based, and consistently proved to be more sensitive to acoustic emissions originating in the specimen than any of the waveguides shown in Figure 3.

The AE signal sampling and processing procedure is illustrated schematically in Figure 2.

Distribution Analysis

Amplitude Distribution. Amplitude distribution analysis is an important mode of AE signal processing and probably the most widely accepted form of distribution analysis. The Dunegan/Endevco analyzer shows a transduced signal peak amplitude. A linear AE signal is converted to logarithmic form thereby creating a wide dynamic range. The log amplified signal is peak detected and an analog to digital converter then outputs a series of 0 to 100 pulse that represent peak amplitude in decibels, with 1 dB resolution.

The peak amplitude distribution function, E (Vo), is the number of events with amplitudes exceeding a given voltage value, Vo. It is of the form:

FIGURE 3

WAVEGUIDE CONFIGURATIONS

Standard End Mount Design

Dunegan/Endevco End Mount Design

Primary Side Mount Waveguide

Revised Side Mount Design

FIGURE 4

LOAD TRAIN CONFIGURATION FOR IN-100 LCF STRAIN CONTROL TESTING ILLUSTRATING
SENSOR PLACEMENT

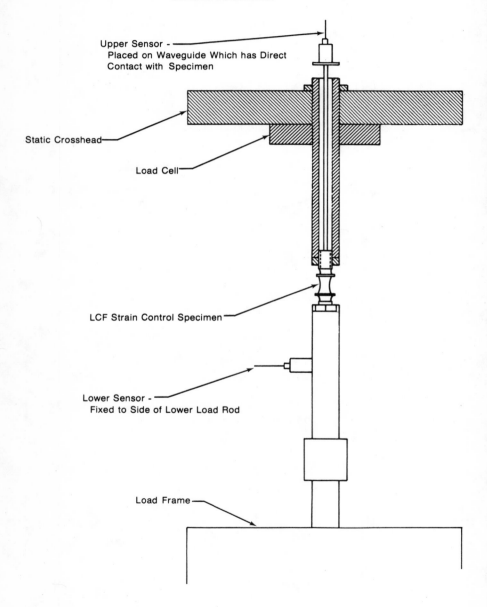

Upper Sensor -
Placed on Waveguide Which has Direct
Contact with Specimen

Static Crosshead

Load Cell

LCF Strain Control Specimen

Lower Sensor -
Fixed to Side of Lower Load Rod

Load Frame

$$E(V_o) = KV_o^{-b}, \text{ where} \tag{1}$$

K and b are constants for a given increment of loading on a structure.

The exponent, b, is sensitive to changing distribution patterns and has therefore been used as an indicator of general structural integrity.

In the logic of the distribution analyzer, each pulse represents the ratio of the input signal to $100\mu V$ (in decibels) through the relation:

$$\text{PULSE} = 20 \log \frac{V_{in}}{100\,\mu\,V} \tag{2}$$

Pulse Width Distribution. The pulse width distribution analyzer counts the time (10μ sec resolution) that an envelope processor is energized, i.e., pulse width per event. An event is defined when the processor is energized at the first pulse of a burst. It remains energized during the burst and stays energized for a fixed envelope time after the last pulse of the burst above threshold. The end of the envelope time is defined as the end of an event. Total time the processor is energized is the burst pulse width.

Ringdown Counts Per Event. The ringdown counts distribution analyzer monitors the number of threshold crossings in each acoustic emission event. For a perfectly resonant transducer with a narrow frequency band, ringdown counts per event, pulse width, and peak amplitude would all be directly related when monitoring rapid-decaying burst-type acoustic emission. The transducers used during this program were relatively sensitive (within 5 dB) over a 300 KHz bandwidth, having a resonant frequency near 650 KHz. Though the burst-type acoustic emission monitored during low cycle fatigue testing displayed exponential decay in many cases, the three distributions functioned somewhat independently with amplitude being the most autonomous.

<u>Analyzer Operation</u>. A processed AE signal which enters the dis-
tribution analyzer is sorted and assigned to 1 of 101 event slots,
according to the operational mode and calibration of the analyzer.
Slot values are represented along the analyzer output X axis,
while the number of AE processed events sorted to each slot is
represented along the output Y axis (Figure 5). A logarithmic Y
axis output was selected in order to accentuate the first few AE
event registrations in each slot.

The Dunegan/Endevco distribution analyzer Y output may be
summed from higher slot values to lower values (Figure 6) and an
average slope, "b", of the summed function determined by the analyzer. This slope may be output to a strip chart recorder and
plotted vs time or test cycles.

The AE logarithmic/summed slope was used periodically during
this test program as an indicator of changing distribution patterns.
Though the log/sum slope does not reflect the total number of
accumulated AE events, it provides a continuous monitor of changes
in AE signal pulse width, ringdown counts (per event) or amplitude
patterns. Since the analyzer sums from higher to lower slot values,
a decrease in slope indicates a higher percentage of high value AE
events. No slope change of course indicates the AE signal distribution being monitored is remaining stable(2).

(2) Pollock, A. A., "Acoustic Emission Amplitudes," pp 264-269,
     <u>Nondestructive Testing</u>, October 1973.

Y Axis:
Number of Events At Each
Slot Value Slot

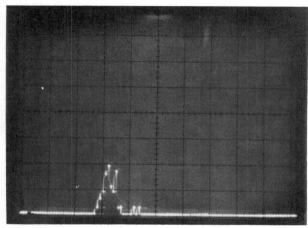

Y Axis May Be Selected
As A Logarithmic Scale

X Axis: Slot Values

0 Slot   100 Slot

Cumulative Slot Values

FIGURE 5   Distribution analyzer histogram normal output

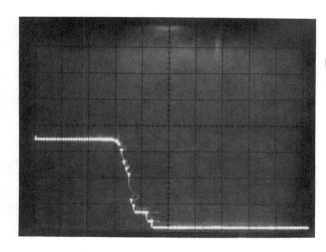

Histogram Y Values
Are Summed From
Larger To Smaller
X Values
(From Right To Left)

Summation

FIGURE 6   Distribution analyzer histogram summed output

TEST RESULTS

Twenty-two LCF strain control specimens from five different heats of IN-100 were tested while monitoring with AE multiparameter distributions. Specimen test conditions, cycles to failure and other pertinent data are listed in Table I.

TABLE I - LCF STRAIN CONTROL TESTING OF IN-100

Specimen Number	Material Source	Engine LCF Cycles	Test Strain Level* (%)	Test Temp. (°F)	Test Cycles	Remarks
217-1	Disk FX 217**	1,761	1.0	1000	9,984	Macrocracked.
217-2	Disk FX 217	1,761	1.0	1000	4,559	Fractured.
217-3	Disk FX 217	1,761	1.0	1000	310	Specimen overstrained. Test discontinued
217-4	Disk FX 217	1,761	1.0	1000	4,200	Test stopped for microcrack inspection.
SC-1	Disk BDB 257**	1,123	1.0	1000	2,279	Macrocracked.
SC-3	Disk BDB 257	1,123	1.0	1000	3,356	Fractured.
036-1	Disk PO36**	1,001	1.0	1000	3,811	Test stopped for microcrack inspection.
036-2	Disk PO36	1,001	1.0	1000	9,974	Fractured.
036-3	Disk PO36	1,001	1.0	1000	2,225	Fractured.
036-4	Disk PO36	1,001	1.0	1000	1,670	Macrocracked.
V-7	Virgin Forging	None	1.0	1000	3,482	Fractured.
V-8	Virgin Forging	None	1.2	1000	1,967	Fractured.
V-9	Virgin Forging	None	1.0	1000	3,837	Macrocracked.
V-10	Virgin Forging	None	1.0	1000	6,027	Fractured.
V-11	Virgin Forging	None	1.0	1000	9,574	Fractured.
V-12	Virgin Forging	None	1.0	1200	2,647	Fractured.
V-13	Virgin Forging	None	1.0	1000	5,310	Fractured.
V-14	Virgin Forging	None	1.0	1000	2,700	Fractured.
V-15	Virgin Forging	None	1.0	1000	5,460	Fractured.
V-17	Virgin Forging	None	0.8	1200	8,845	Test stopped for microcrack inspection.
V-21	Virgin Forging	None	1.0	1000	3,325	Fractured.
V-22	Virgin Forging	None	1.0	1000	8,487	Fractured.

* Mean Strain = ½ Maximum Strain.
** Engine-cycled material.

Typically, acoustic activity and distribution analyses during the course of an LCF test may be modeled as:

1. A period of initial shakedown. AE rate registry is high. Pulse width, ringdown counts per event, and amplitude distributions reflect AE bursts with greatly varying characteristics.

2. A stable period. AE rate decreases. AE distributions indicate bursts with relatively short pulse widths, few ringdown counts per event, and low peak amplitudes.

3. Maximum AE signal pulse widths and maximum ringdown counts per event suddenly increase (possibly several times during this portion of the test), but AE signal peak amplitudes change relatively little. AE rate may slightly increase.

4. Another stable period. AE distributions again indicate bursts with relatively short pulse widths, fewer ringdown counts per event and unchanged signal peak amplitudes. AE rate is low.

5. A general increase in AE activity to specimen failure. AE rate may increase two or more orders of magnitude. AE distributions indicate burst registries with longer pulse widths, increased ringdown counts per event and higher peak amplitudes.

The above phenomena are represented graphically in Figure 7.

Testing of three specimens (036-1, V-17, and 217-4) was discontinued during period "3" for surface and internal inspections of the gage sections. External inspections were performed using eddy current and fluorescent penetrant techniques while the internal inspection procedure included sectioning each specimen axially and polishing through the inspection surface while microscopically scanning at 100X to 1600X. Thin-foil microscopy and replication at 20,000X were also performed on V-17.

Eddy current and fluorescent penetrant inspections of specimen 036-1's gage section surface produced no fatigue crack indications. However, upon sectioning the specimen axially and subjecting it to microscopic fractographic examination, multiple microcracks (on the order of 1/4 mil) were discovered internal to the specimen gage section, initiating primarily from microporosity. Further examination outside the most highly strained region of the gage section revealed no microcracking in the porous areas, discounting the possibility that the material was microcracked prior to testing.

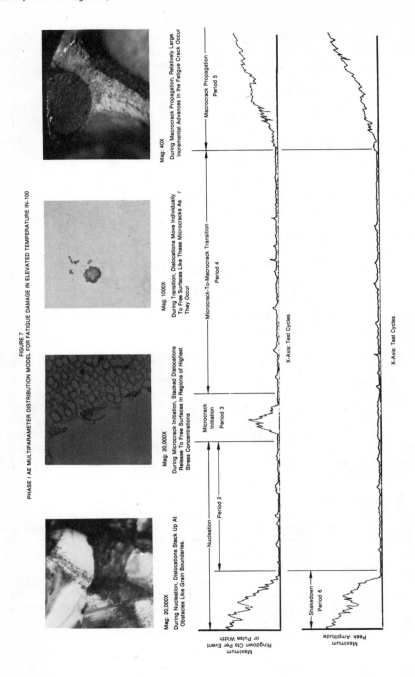

FIGURE 7
PHASE I AE MULTIPARAMETER DISTRIBUTION MODEL FOR FATIGUE DAMAGE IN ELEVATED TEMPERATURE IN-100

Mag: 20,000X
During Nucleation, Dislocations Stack Up At Obstacles Like Grain Boundaries.

Mag: 20,000X
During Microcrack Initiation, Stacked Dislocations Release To Free Surfaces In Regions of Highest Stress Concentrations

Mag: 1000X
During Transition, Dislocations Move Individually To Free Surfaces Like These Microcracks As They Occur

Mag: 40X
During Macrocrack Propagation, Relatively Large Incremental Advances In the Fatigue Crack Occur

Nucleation — Period 2
Microcrack Initiation — Period 3
Microcrack-To-Macrocrack Transition — Period 4
Macrocrack Propagation — Period 5
Shakedown — Period 6

X-Axis: Test Cycles
X-Axis: Test Cycles

Maximum Ringdown Cts Per Event or Pulse Width
Maximum Peak Amplitude

Testing of specimen 217-4 was also discontinued after the period "3" of increased AE signal maximum pulse widths and ring-down counts per event (Figure 8). Though 217-4 was machined from a different heat of material than 036-1, post-test external and internal inspections produced similar results. Gage section internal inspection again revealed microcracking from microporous areas at the center of the gage section while no microcracking was noted outside the most highly strained region of the specimen.

Replicas from the V-17 gage section showed multiple micro-cracks adjacent to the primary gamma prime particles. Thin-foil microscopy revealed numerous stacked dislocations at particle boundaries as well as dislocation slipbands throughout the highly strained material.

Four tests (217-1, SC-1, 036-4, and V-9) were discontinued near the beginning of AE period "5" (general increase in AE activity). Eddy current inspection disclosed macrocracks in all specimen gage sections. The distribution records of maximum pulse width, ringdown counts per event, and peak amplitude for the 217-1 test are presented in Figure 9, illustrating periods "1" through "5"

Tests which did not exhibit the above AE phenomena (periods "1" through "5") encountered long periods wherein AE registry was completely masked by extraneous noise, or multiple macrocracks were obvious on the specimen fracture surfaces, indicating a combination of mechanisms operating simultaneously (Figure 10).

CONCLUSIONS

It is apparent from the test results that at least four different stages of fatigue damage in elevated temperature IN-100 may be characterized using time-domain AE multiparameter distributions. The respective AE characteristics are illustrated in Figure 7. The four fatigue stages are:

Stage 1 - Nucleation

1. MPD Phenomena

    - All three time domain distributions (ringdown counts per event, pulse width, and peak amplitude) fluctuate during the initial "shakedown" portion of a typical test, followed by relatively small consistent parametric values.

FIGURE 8

LCF SPECIMEN 217-4, $\Delta\epsilon$ = 1.0%, TEMP = 1000°F
TEST DISCONTINUED WHEN AE MPD INDICATED MICROCRACKING (PERIOD 3)

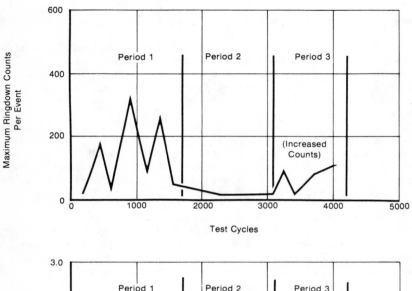

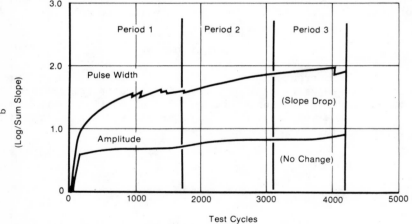

Decreasing "b" Means a Higher Percentage of Larger Value Pulse Widths or Amplitudes

FIGURE 9

LCF SPECIMEN 217-1, $\Delta\epsilon$ = 1.0%, TEMP = 1000°F
TEST TERMINATED WHEN ACOUSTIC OUTPUT INDICATED
MACROCRACKING (PERIOD 5)

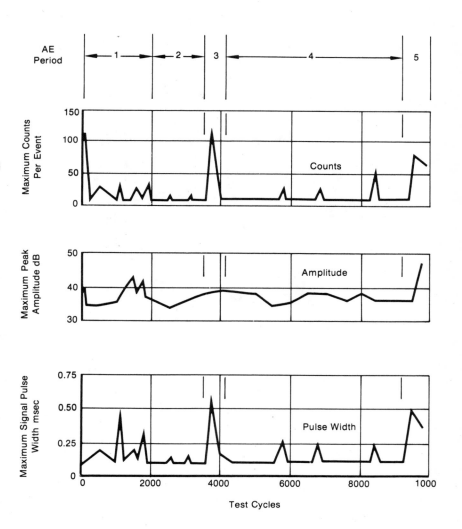

FIGURE 10
LCF STRAIN CONTROL SPECIMEN V 18
STRAIN RANGE: 1.0%, TEMP = 1000°F
FRACTURED AFTER MULTIPLE MACROCRACK INDICATIONS

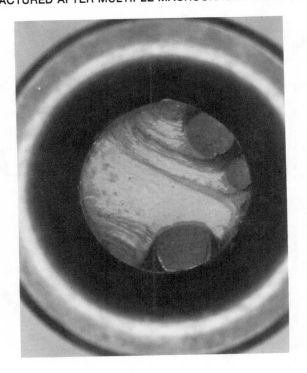

2. Mechanism

- "Shakedown" is a combination of extraneous noise from the test load train and acoustic output from initial plasticity at the specimen gage section.

- During nucleation, dislocations stack up at obstacles such as microporosity and grain boundaries.

Stage II - Microcrack Initiation

1. MPD Phenomena

- Increased maximum ringdown counts per event,

- Longer signal pulse widths, and

- Little variation in peak amplitudes.

2. Mechanism

- Stacked dislocations begin to move to free surfaces in regions of highest stress concentrations both internally and externally. These events occur individually over longer periods of time than stackup (producing longer AE signal pulse widths and greater ringdown counts per event) but energy release is not great enough to appreciably affect AE amplitude distributions.

Stage III - Transition

1. MPD Phenomena

- Maximum ringdown counts per event and maximum signal pulse widths decrease to levels occurring previously during the nucleation stage.

- Amplitude distributions still display little change.

2. Mechanism

    • Dislocations are moving to free surfaces
      (microcracks) as they occur. Lack of
      stacking before dislocations are released
      produces shorter lived individual acoustic
      bursts (i.e., shorter pulse widths and
      fewer ringdown counts per event than during
      microcrack initiation).

    Stage IV - Macrocrack Propagation

1. MPD Phenomena

    • All three time domain distributions display
      increases in maximum parametric values.

2. Mechanism

    • Large incremental advances in the fatigue
      crack(s) occur, producing relatively large
      acoustic energy release with each advance.
      Internal crack propagation reaches the
      specimen surface and oxidation layer crack-
      ing (as well as the large acoustic energy
      release) contributes to higher acoustic
      peak amplitudes. Each event occurs over a
      longer period of time than those occurring
      during nucleation or transition phases,
      thus producing longer signal pulse widths
      and more ringdown counts per event.

Detection of fatigue macrocracking may be performed solely
on the basis of AE peak amplitude distribution analysis. The
results support the theory(2) that larger deformation events
give a larger relative yield of acoustic emission energy and,
therefore, larger AE peak amplitudes.

A problem of AE calibration is involved when conclusions
from the strain control program are applied to a practical test,
such as screen testing of turbine engine disks while monitoring
with AE techniques. All parameters discussed in this text were
left in relative terms which make application from one test
form to another convenient, providing that the subject specimen
has no initial fatigue damage. However, when applying the
principles of multiparameter distribution analyses to a screen

test in which very few load cycles will be applied to the specimen and the objective of AE monitoring is to determine the highest degree of fatigue damage incurred in the specimen, it is necessary to deal with parameter absolute values. Current studies at P&WA include transducer calibration schemes and, while a satisfactory method has not been developed, there are several promising concepts in use throughout the industry (3, 4).

(3) Hsu, N. N., "A Mechanical AE Simulator For System Calibration and Waveform Analysis," National Bureau of Standards, presented at 16th Meeting of AEWG, Williamsburg, Va., October 1976.

(4) Feng, C. C., "Acoustic Emission Transducer Calibration Progress Report," Endevco Corporation, presented at 16th Meeting of AEWG, Williamsburg, Virginia, October 1976.

QUANTITATIVE MEASUREMENT OF THE RELIABILITY OF NONDESTRUCTIVE INSPECTION ON AIRCRAFT STRUCTURES

W. H. Lewis, W. H. Sproat, and W. M. Pless
Lockheed-Georgia Co.
Marietta, Georgia

and

B. W. Boisvert
Air Force Logistics Command
Kelly AFB, San Antonio, Tx.

ABSTRACT

The materials and designs of our current aircraft do not tolerate large crack-like flaws. Critical, highly stressed components in some of the newer aircraft cannot tolerate even a one-half-inch long crack and remain safe for further flight. Repair costs are also severely impacted by the detectable flaw size. Those responsible for the safety and integrity of today's aircraft are continually requiring detection of smaller and smaller defects and the inspection of larger and larger areas with a far greater confidence in the inspection results than ever before. The fact is, however, that the reliability of nondestructive methods to detect cracks during a maintenance inspection of aircraft structure has never been measured.

This paper describes in detail a U.S. Air Force-funded program currently underway to quantitatively measure the reliability of nondestructive inspection (NDI) conducted on actual aircraft

structure at Air Force field and depot installations. The problems involved in quantitative
measurement of the reliability of NDI, as performed in the maintenance inspection of aircraft,
are discussed. A detailed approach is presented
for the collection of flaw detection data,
including sample selection and preparation,
inspection procedures development and standardization, and the logistics involved in obtaining
a statistically representative cross section of Air
Force field and depot installations. The types
and quantity of data being collected are reviewed
along with approaches for analysis of the variables
involved in the program.

INTRODUCTION

The high-strength materials and efficient designs of our current
aircraft do not tolerate large crack-like flaws. Critical, highly
stressed components in some of the newer aircraft cannot tolerate
even a one-half inch long crack and still remain safe for further
flight. Repair costs are also severely impacted by the detectable
flaw size, smaller cracks being significantly less expensive to repair than larger cracks. Those responsible for the operational
maintenance and integrity and safety of today's aircraft are
continually requiring detection of smaller and smaller defects
and the inspection of larger and larger areas with a far greater
confidence in the inspection results than ever before. NDI
technicians are being asked to precisely find a small isolated defect
during the inspection of a large number of potential defect sites.

An interesting question is: Can NDT provide the increased sensitivity with higher levels of confidence now being demanded of it?
The fact is that in most cases we really don't know the reliability
of nondestructive methods to detect flaws during routine inspections.
Within the aerospace industry, specification changes during the
last several years have prompted requirements for contractors to
demonstrate their ability to detect harmful size defects. Recent
NDT Reliability Demonstration Programs conducted by several major
aerospace manufacturers have given us an idea of how good we can
expect our production inspection to do in finding various size
defects. But this is in a production environment only and most of
these demonstrations have been conducted under controlled conditions. A maintenance or overhaul environment is completely

different than a production situation, and we don't have much data
on how good NDT is under these field conditions.

It is safe to say, then, that we don't know how well or how badly
our NDT has always performed for us in the past. We are only
certain about the flaws we do find and verify. But for the structure
for which no flaws are reported, we cannot be sure that flaws have
not been missed.

It became apparent in the late 1960's that the state-of-the-art
capabilities and practices in NDT were not sufficient to meet the
needs of the emerging damage tolerance design philosophies which
take into account inevitable defects in all materials. NDT not only
failed to provide the quantitative data about flaws that fracture
mechanics required, but it was seen to be incapable of always
finding all of the flaws of subcritical size. And, we openly admit,
cracks of even larger size can be missed.

Crack detection was felt to have a statistical nature, with the
smaller cracks having a low probability of detection and the larger
cracks having a probability of detection approaching a maximum
level somewhat below 100 percent. We will now look at some of
the problems and approaches in quantitatively measuring the
reliability of NDT to find various size flaws and the specifics
regarding the initiation and status of one such effort currently underway by the U. S. Air Force.

DISCUSSION

Problems In Quantitative Measurement of NDI Reliability In Maintenance Inspection of Aircraft

Along with the new damage tolerant design concepts, structural
analysis technology has improved. Through the use of fracture
mechanics technology we can predict the rate at which a crack-
like defect will propagate under known loading conditions and
thereby formulate analytical projections of service life as shown in
Figure 1. The one thing we don't know now is how good are in-
service inspections. Where on this curve should inspections com-
mence? And, how often should inspection intervals be planned?
Where, exactly, on this curve would we feel confident that NDI
would reliably find a given crack?

TYPICAL EXAMPLE OF PREDICTED
CRACK GROWTH

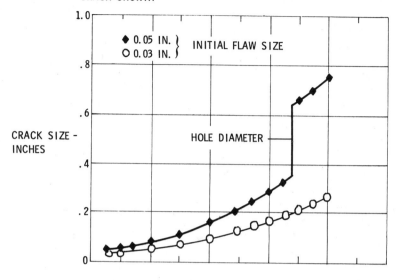

FLIGHT HOURS

FIG. 1. CRACK SIZE VS FLIGHT TIME

There are very little data to substantiate the capability of NDI
under aircraft maintenance and overhaul conditions. All that
exists now are estimates and we're not sure how good these
estimates really are. No measurements have been made on the
reliability of NDI under actual field and depot conditions.
Measurements have been made under laboratory and manufacturing
environments, which are entirely different than a maintenance
environment. Ability to reliably detect small defects would like-
wise be quite different.

The problem of determining the reliability of nondestructive
inspections in a maintenance environment is extremely complex.
There are a number of closely interrelated factors that dramatically
affect the reliability of NDI. For example, if a single part is
brought into the laboratory and it contains a flaw whose locations
and orientation is known, it is a relatively simple matter for the
NDI technician to demonstrate the flaw's presence even though

quite small. This he will be able to do with a high degree of
reliability. Current equipment is capable of detecting extremely
small flaws under ideal conditions. It is however, a different
matter entirely if 50, 500, or even 5000 parts were to be inspected
and it is not certain that any of them contain a flaw.

The situation is further compounded when the inspection is removed
from ideal laboratory conditions to some of the more adverse
maintenance and overhaul environments that exist at some locations.

The Lockheed/AFLC NDI Reliability Program

General Description

Faced with the fact that nondestructive inspection reliability was
an unknown factor at a time when airframe design philosophies and
military specifications were requiring increasingly more accurate
and sensitive quantitative NDT data, the Air Force Logistics
Command contracted with the Lockheed-Georgia Company in June
1974, to institute a program for assessing Air Force NDI performance
during maintenance and overhaul operations.

The purpose of the program, which is now underway, is to determine
the actual capabilities of nondestructive inspection for detecting
various size service-induced defects in actual aircraft structure
under true field and depot conditions at Air Force installations.
In-service NDI capabilities are extremely important to the
implementation of structural life extension and damage tolerant
design practices which are based on the ability to detect flaws of
specified sizes and to predict flaw growth under service conditions.
The philosophy of this program centers around the measurement and
recording of a number of uncontrolled variables as well as the
treatment of a select group of controlled variables. It was designed
to passively measure the field and depot flaw detection capabilities
and let the chips fall where they may. The rigorous controls
normally applied in laboratory evaluations have been avoided. The
data sample size and method of treatment, however, will allow for
attaching statistical significance to the observations of both controlled and uncontrolled variables.

The approach to determine the reliability of NDI is to take typical
aircraft structure with known cracks directly to the field and depot
NDI shops, where the normally assigned personnel are asked to conduct routine nondestructive inspections on the pieces of hardware.
The key point is that actual aircraft structure with service-induced
fatigue cracks are circulated to typical aircraft maintenance
installations and inspected by personnel and equipment normally
used to do these jobs. The results of the NDI are examined in terms
of flaw detection probabilities as a function of flaw size on a
statistical basis.

At each installation visited, the personnel of the participating NDI
shop are given an identical orientation briefing in which the details
and goals of the program are explained. The data to be accummu-
lated and all facets of his participation are described. Technicians
are then assigned specific NDI tasks on the samples. The program
is fully described in Reference I, which also includes test sample
descriptions and the detailed nondestructive inspection procedures.

Transporting The Specimens

The project is being accomplished by bringing to each installation
a utility trailer equipped with all the structural samples and other
accessory equipment needed to support the entire program. The
trailer arrives directly at the installation, is set up in a relatively
self-sufficient manner in a location somewhat out of the way not to
interfere with normal maintenance activities, and the NDI technicians go about routinely inspecting the samples. The trailer also
serves as an object on which to mount samples for radiographic and
eddy current NDI.

Structural Samples

Six types of representative structure comprise the array of test
samples (I). These are structural pieces that have been removed from
actual in-service aircraft or aircraft test structures. The samples are
representative of wing boxes, skin and stringer segments, wing panel
risers, wing panel segments, forged fittings and spar box assemblies.

One of the major problems encountered in originating the program
was that of obtaining a group of test samples which collectively

present a distribution of crack sizes ranging from very small ($\leq$ 0.050-inch) to approximately one inch. Most cracked material is repaired or removed from an airframe when the crack is in the 1/4 to 1/2 inch range. This limited range would not permit an adequate evaluation of flaw detection capability relative to a representative range of flaw sizes. It was necessary, therefore, to examine a considerable number of available structures and pieces removed from service to select suitable samples for the program.

One of the samples is an intact 5-foot segment of an actual center wing box shown in Figure 2. This all-aluminum sample contains catalogued fatigue cracks on the lower surface oriented chordwise at fastener holes, drain holes and cut-outs. The original wing box had received service-induced fatigue damage, was removed from service and cyclically loaded in a test jig to further induce fatigue damage prior to residual strength testing. The segment is now mounted on a dolly with the lower surface facing upward for performance of NDI.

Another series of samples consists of eight segments of a typical wing lower surface having a history identical to the previous large sample. These segments consist of aluminum skin and hat-section stringers and have typical dimensions of 16" X 20". The samples can be mounted in various positions in the trailer to simulate overhead or other inspection situations. One of these can be seen in Figure 3 mounted in the trailer for radiographic NDI.

Another series of samples are simulated portions of an engine pylon aft truss. They were cycled in a fatigue machine to induce small fatigue cracks along the edge. These samples have dimensions of 2.5" X 18" and are made of titanium for ease of cleaning between repeated penetrant inspections. For inspection, they are nested in one end of the wing box segment to simulate wing riser stiffeners.

A fourth series of samples are portions of another type of aluminum surface panel containing real fatigue cracks at fastener holes.

A fifth series of samples is composed of bare segments of forged aluminum wing fittings with service-induced fatigue cracks at fastener holes. These small segments are attached to a single cover plate to provide a layered stack-up, typical of multi-layered joints.

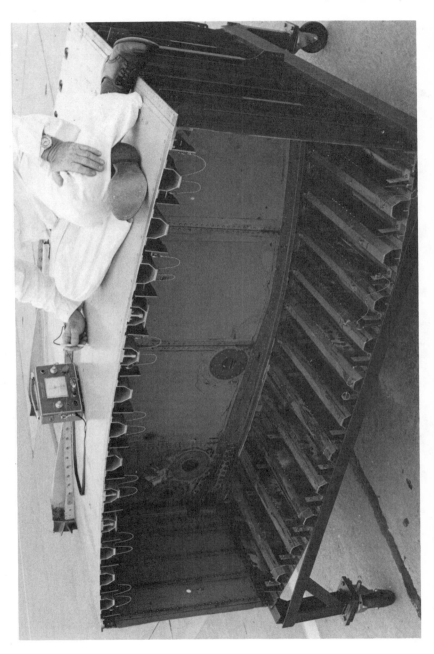

FIG. 2.   WING BOX SPECIMEN

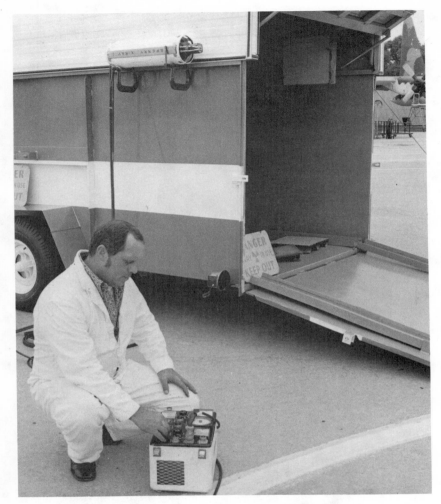

FIG. 3.   WING SPECIMENS MOUNTED IN TRAILOR FOR X-RAY NDI

The final structural type is a box beam structure designed to
represent wing spar cap-to-web assembly joints and web-to-
stiffener attachments. The box beam shown in Figure 4 was fatigue
tested for 147,000 cyclic test hours which produced a number of
fatigue cracks at fastener sites. This aluminum structure is also
mounted on a dolly for handling ease.

FIG. 4.   SPLICE IN BOX BEAM SPECIMEN

Program NDI Techniques And Procedures

The NDI techniques being evaluated in the program are those that
are typically used at both field and depot levels. [1] These include
eddy current inspection, both surface scanning around the periphery
of a fastener head and bolt hole eddy current inspection with the
fastener removed. Ultrasonic inspection is included for fatigue
cracks progressing out from under the head of a fastener. Radiographic inspection capabilities are evaluated by using the transport trailer to simulate a section of aircraft fuselage structure
containing the test sample. An evaluation of penetrant inspection
capabilities is included in a set-up simulating the interior of a
wing box. With the exception of magnetic particle inspection,
these NDI methods are those that are most currently used in our
everyday maintenance inspection. During the planning phase of
the program, an analysis determined that the cost of obtaining the

large number of steel samples containing service type defects was prohibitive in both time and dollars, and they were therefore omitted from the project.

Recently, automated versions of an ultrasonic shear wave scanner and an eddy current bolt-hole scanner were incorporated into the program for evaluation at the Air Force installations remaining to be visited.

Detailed "how-to" procedures were written for the NDI techniques to be used on each sample type. The procedures follow the requirements of MIL-M-38780A and contain all the information necessary to instruct trained technicians how to perform the desired inspection. This information includes equipment selection, equipment set-up and calibration and performance of the inspection. Detailed illustrations are included showing the sample, areas to be inspected and steps to be followed. All NDI procedures, including calibration procedures, were incorporated into a manual and are made available to the participating technicians.

Calibration standards for the ultrasonic and eddy current procedures are furnished. A step wedge is provided for developing log relative exposure curves for the particular X-ray equipment. The standards and the manual assure uniform inspection of each sample. Checks on the condition and status of equipment are part of the essential data which is compiled.

Steering Committee

A Steering Committee of recognized government and industry experts in nondestructive inspection technology was established to review this program prior to initiation and provide overall guidance and direction. In the fall of 1974, this committee met and pointed out the need to expand the types and quantity of samples to obtain a wider variety of aircraft structure than originally planned. They also recommended that the number of flaws in each size range be increased to provide a totally realistic judgement of true NDI reliability. Because of the variations between various Air Force Commands and their differences in applying NDI to their operational requirements, the program was expanded to include a broader sampling of field installations.

Air Force Field And Depot Participants

The program is now scheduled to include four to five bases at each
of the four major commands including Air Training Commands,
Military Airlift Commands, Strategic Air Commands, Tactical Air
Commands and all five Air Logistics Centers. With an average of
five technicians participating in the complete program at each base,
and 20 technicians at each depot, the necessary data sample size
will be available to ensure that the results are statistically valid.

Twenty-two Air Force installations are scheduled for visitation and
presently the team is at the 17th installation.

In selection of personnel who will be participating at the bases,
care is taken to select, when possible, a representative cross-section
with regard to training, skill level, and current proficiency. Some
bases have limited numbers of personnel and all available NDI
technicians may have to be used to acquire sufficient quantities of
data.

Data Recording

There are four specific types of data forms [1] that each NDI technician is required to complete as he or she progresses through the
program. One pertains to his training and background, one is a
data sheet for each inspection that the technician conducts, one
pertains to an operational checkout of the various types of NDI
equipment, and the last one is a scale drawing of the sample being
inspected which is used by the technician to record the crack
indications that have been marked on the samples.

The Technician Profile Form is completed by each participating
technician at the beginning of the program and is designed to provide background information on all Air Force NDI Technicians.
Because this program is not an evaluation of inspectors as individuals, names are not used. Each NDI Technician is assigned an
identification number which is used on all data forms completed by
the technician. On this particular data form, we ask for the date,
assigned ID number, job title or AFSC, general educational information and then more specifically, some questions about his or her
training in NDI. This information will enable us to determine if
certain types of background, formal training, prior jobs, and so

forth, seem to have an affect on the accuracy of the inspector's results. Similarly, the number of inspections over a typical month's time period is also recorded in an attempt to assess proficiency. We also ask for certain physical data such as height, weight, age, sex, and so on. This will enable us at the completion of the program to evaluate many of the human factors involved in inspection reliability. For example, do married technicians do a better job than single technicians, on the average; or do older, more experienced technicians really do a better job than younger, less experienced technicians?

The Inspection Data Sheets pertain to the recording of the technical information regarding each individual inspection conducted. This information includes equipment identification, serial numbers, equipment settings, standards, transducers, materials, inspection times, and various other comments that pertain to the particular inspection technique and procedure used on the individual samples.

The Equipment Performance Data Sheets are designed for recording the results of certain specific checks on the NDI equipment used to inspect the samples. This not only gives information regarding crack detection capability as a function of equipment performance, but it will also tell us how well the NDI equipment currently in the inventory is being maintained.

During each inspection, except Radiography, crack indications are marked directly on the structure samples by the technician with special marking instruments provided. At the completion of the inspection, the technician transposes the crack indications marked on the sample directly to a scale drawing of the test sample. The program coordinator double checks each of these sheets against the marked structure sample to ensure that the transposition is accurate before he removes the markings on the sample so that it will be ready for the next technician.

A Facility Evaluation Form is used by the accompanying test engineer to record pertinent details of the location and general area in which the inspections are conducted, and the ambient conditions that exist during the inspections, such as peculiar light or noise levels, and a complete description of the inspection area. Finally, a Daily Log is maintained by the test engineer to record the details of each day's operation, such as any special conditions

that existed, the weather conditions for that particular day, the
total number of technicians available, and the amount of samples
inspected. It also provides a unique opportunity to record the
activities of each technician on a daily basis.

Collection of all these data will provide a unique opportunity to
evaluate the significance of many factors other than just flaw size.
These include equipment performance, the environment, the
effectiveness of procedures, training and background and most
importantly, the numerous human factors involved in the technician's performance.

Some Program Schedule Milestones

Before initiating data acquisition in the field, a trial run was made
at Wright-Patterson Air Force Base in the summer of 1975 to ensure
that this phase of the program would go smoothly. It also enabled
a large number of Air Force scientists and engineers to view this
portion of the program first-hand. We are now more than two-
thirds the way through the planned data acquisition phase. The
only way to tell precisely what cracks in the test samples have been
found and what cracks were missed, and their true size, is to
destructively section each test sample for metallographic measurement. This will be done after all data is acquired. The final report
will be released after sectioning of the test samples is complete to
verify the actual presence and size of each crack.

Some Preliminary Data Trends

At the middle of April, 1977, we had completed visits to twelve
different field installations and two Air Force Logistic Command
Depots. The data collected from these installations are now being
compiled and are beginning to show us some preliminary trends of
NDI capability. We hesitate to publicize preliminary and incomplete data and therefore will wait until all the data are acquired
before publishing any results at Air Force request.

The results at this time do indicate that overall NDI reliability may
be less than some thought. Areas have been indicated for potential
improvement or for changing our traditional approaches. For
example, it has long been assumed that the more formal training a

technician has in the technical aspects of NDI, the more acuity he will have in detection of flaws. But Figure 5 seems to say that formal training beyond a minimum level sufficient to thoroughly acquaint the technician with the background and use of a technique is of questionable value. If this trend turns out to be true, what should be our remedial approach? Do we drop advanced training courses, or do we change their emphasis from technical to motivational aspects?

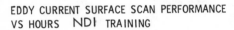

EDDY CURRENT SURFACE SCAN PERFORMANCE
VS HOURS NDI TRAINING

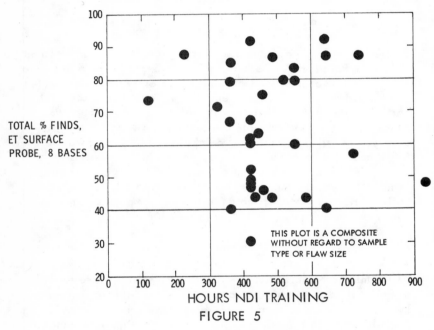

FIGURE 5

Next consider the influence the number of years of experience has on detection ability. Our intuition tells us that the more experience, the better the ability should be. But take a look at Figure 6 which shows a plot of detection rate for eddy current surface scan versus years of experience. No clear-cut influence for experience is seen, except that poor performance is perhaps moderated by experience. Many of those with one to two years of experience performed better than some of those with 6 to 7 years.

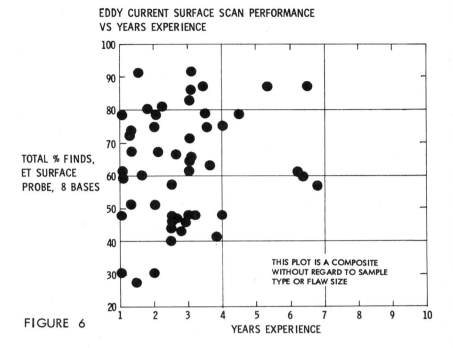

FIGURE 6

Figure 7, suggests that a third factor, inspection frequency or the number of times a technician conducts NDI in terms of assignments per month, exhibits no correlation with performance.

Correlation may or may not exist in other cases and there could be something hidden in the totals of these three cases. For example, the lower 25 percent of the performance could be dropped out and perhaps a correlation would then exist. We don't know right now. The important point here is that there are a large number of variables involved in the conduct and performance of a non-destructive inspection. The data on all these variables are being collected and a thorough analysis and evaluation is now required to determine which of the variables affect the reliability of inspections.

A final example of the preliminary findings is given in Figure 8 which shows a distribution of technician performance with eddy current NDI at ten bases. It shows that most individuals are finding 60 to 70 percent of the known (catalogued) cracks in one

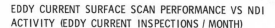

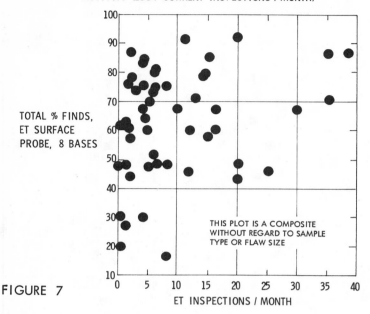

EDDY CURRENT SURFACE SCAN PERFORMANCE VS NDI
ACTIVITY (EDDY CURRENT INSPECTIONS / MONTH)

FIGURE 7

test sample. The obvious goal is to have the peak in the 90 to 100 percent range.

CONCLUDING STATEMENTS

I hope this presentation has given you an idea of what we are doing on this important program and what we hope to learn. Most important is what our final analyses will indicate about NDI as now practiced and how will we all react to the findings. Although this particular program is designed specifically at measuring the effectiveness of Air Force nondestructive inspections, the results certainly will have impact on most other industries that frequently rely on in-service inspections to ensure structural integrity. The lessons we will learn here will have far reaching affects for many years to come.

One alternative, which is not necessarily advocated here, is to conclude that what we have measured is about as good as can be expected for existing NDI capabilities. If this is the case, then

DISTRIBUTION OF TECHNICIAN PERFORMANCE
AT 10 BASES, SAMPLE A, EDDY CURRENT
SURFACE SCANS AROUND FASTENERS

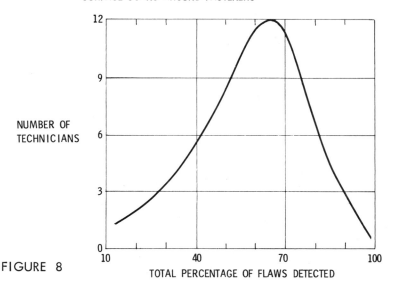

FIGURE 8

inspection intervals, assignment of redundant inspections, and safe-life limits should be established for these data. If one-half inch is the smallest crack size we can reliably expect to find with existing capabilities, accept it as such and adjust maintenance operations to accommodate the situation. Similarly, designs of new aircraft would be aimed at tolerating something greater than one-half inch cracks. This alternative of accepting the present NDI capability,however, is not economically practicable.

A cursory look at our data so far shows that the ability to reliably detect cracks has very little relationship to the type or the amount of formal NDI training a technician has received. This implies that except for a certain amount of general baseline training in NDI that additional training now conducted is not effective in increasing the technician's skill or proficiency level as previously shown. We have found, however, and most technicians have indicated to us, that this program has been a unique learning experience for them: the fact that we have been able to bring a wide variety of crack samples to their facility and they have the opportunity to sit down with good straightforward inspection procedures, calibrate their equipment, and do repeated inspections on these samples containing actual service cracks. It has been of tremendous benefit especially to the younger, less-experienced technicians. Perhaps more or better training along these practical lines would show a marked improvement in inspection reliability.

Another area of potential improvement lies in the elimination of very poor inspectors. In a close look at the data the very poor inspectors in many cases, tend to pull the whole average down considerably. The elimination of these poor individual inspectors by some form of practical certification would considerably improve the average reliability.

Optimizing the size of the task is another potentially beneficial approach to improving flaw detection capability. For example, assigning a technician 2,000 fastener sites to inspect is a psychologically overwhelming task, while assigning inspection of only two dozen fastener holes at a time is much more acceptable. The individually assigned task size must be small enough to accommodate the sustained level of vigilance required for a high degree of flaw detection, yet be large enough to be economically practical.

The human factor aspects of inspection reliability are becoming more and more apparent the further we proceed with the program. It is obvious that some people make better NDI technicians than others who seemingly have the same educational background and intelligence level. A somewhat similar problem existed a number of years ago concerning the selection of operators to man radar scopes. During WWII several studies were undertaken to define the human factors involved in the job performance of Radar Operators and, as a result, psychological testing and screening were developed to select those individuals best suited for this particular job. Perhaps a parallel exists here with nondestructive inspection. After all, we do require the NDI technician to perform somewhat monotonous, repetitive tasks, but yet require him to be ever-diligent and alert to detect the slightest defect.

One further question we have concerns the alternative methods of presenting the composite data. There are several ways of showing flaw detection versus flaw size, or technician performance versus flaw size, or technician performance versus any other test variable, and other conclusive relationships. The methods include plotted curves with confidence limits, histograms, and perhaps other graphical methods. What would you suggest as the most useful and instructive means to portray the important findings of this program? We welcome your comments.

In conclusion, the results of the program to-date indicate that the reliability of nondestructive inspections conducted on aircraft may require some improvement if we are to make maximum use of prudent inspections to extend the useful service life of existing aircraft fleets. Some provocative questions are raised regarding where those improvements might come from and how much improvement might be possible. At the end of this program, we hope to have at least a few of the needed answers.

REFERENCES

(1)   Test Plan (for) Reliability of Nondestructive Inspection on Aircraft Structure, AFLC Contract F41608-73-D-2850 P0038, July 1975.

INSPECTION OF EXISTING STRUCTURES

Stanley Gordon
Federal Highway Administration

At the present time, this country is engaged in one of the most important tasks ever undertaken by the American people and its government - that of inspecting our Nation's bridges and rehabilitating or replacing those which are inadequate to handle today's traffic with its increased volume and heavier loadings. We forged ahead building thousands of miles of highways and thousands of bridges; while failing to look back at the deteriorated old structures that lie in our wake.

The catastrophic failure of the Silver Bridge (Figure 1) at Point Pleasant, West Virginia, in 1967 was the catalyst that brought this problem to the Nation's attention. As a result of this catastrophy, the National Bridge Inspection Standards were developed under the 1968 Federal-Aid Highway Act requiring all States to set up a bridge inspection organization capable of performing inspections, preparing reports and rating the bridges on the Federal-aid highway systems in accordance with the provisions of the AASHTO manual and the standards contained therein. It also required that each bridge be inspected at regular intervals not to exceed 2 years, established qualifications for inspection personnel, required an inspection report for each bridge, and required each State to prepare and maintain an inventory of all bridge structures subject to the standards.

Figure 1: Collapse of the Silver Bridge at
Point Pleasant, West Virginia

Realizing that there may be those who are not directly
connected with bridge inspections, inventories, ratings and
replacement, I think it would be appropriate at this point to
briefly give you a status report explaining the national
picture. Currently, there are an estimated 560,000 bridges
in the United States and about 125 to 150 of these collapse
each year for various reasons and an additional number fail
because of floods. We estimate that about 105,500 of these
bridges are either structurally deficient or functionally
obsolete (a structurally deficient bridge is one that has been
restricted to light vehicles only or closed, while a
functionally obsolete bridge is identified as one whose deck
geometry, clearances, approach roadway alignment or load
carrying capacity can no longer satisfactorily service the
system of which it is an integral part). About 33,500 of
these bridges are on the Federal-aid highway system (6,900
structurally deficient and 26,600 functionally obsolete).
While most of the bridges on the Federal-aid system have been
inspected, it is clearly evident that the need to inspect the
remaining 72,000 should be a high priority on anyone's list
and a necessity that cannot be ignored. It should be noted.
however, that many States have had their own inspection program
for not only those bridges on the Federal-aid system, but also
for those bridges under State and local agencies.

The main purpose of any bridge inspection is to assure the
owner that the structure is functioning as intended. It is only
through the process of inventory, inspection and classification
that deterioration, cracking and malfunctioning members can be
properly identified and evaluated to reflect the true condition
of the structure.

While each structure may require a slightly different approach, it is very important to plan the inspection in a systematic manner. One of the more important aspects of the bridge inspection is that of recording information and preparing sketches in the bridge field book. This is particularly important on large trusses where each panel point is assigned a specific designation in order to keep track of the condition of individual members during the inspection and in the office at a later date, when doing the evaluation. Equipment necessary to inspect most of the routine bridges can generally be carried by car or truck (Figure 2); however, for

Figure 2:   Typical Inspection Vehicle

those structures that do not have catwalks or handrails, it may be necessary to purchase or rent a snooper truck (Figure 3) to have the freedom and accessibility needed to do a thorough inspection. In addition, photographs should always be taken to show the existing conditions.

A bridge is essentially made up of two basic components - substructure and superstructure. For the purpose of this discussion, the following comments concerning items to be inspected will only be associated with steel superstructures. We will look at the various components, why that component is significant, and the possible adverse effects that its condition would have on the bridge superstructure as a whole.

Figure 3:   Snooper Truck for Superstructure Inspection

The first area of concern for inspection in the superstructure is the bearings which transmit the weight of the
bridge and the superimposed loads to the piers or abutments.
Bearings are generally classified as being either fixed or
expansion. Fixed bearings do not permit longitudinal movement (expansion and contraction caused by temperature change);
however, they do permit rotation at the bearing point due to
deflection of the span caused by live loads on the structure.
Expansion bearings, on the other hand, do permit both rotation
and translation. While bearings are made from several
different materials, each may be affected by any one of the
following conditions (Figure 4):

Figure 4:   Deterioration, Dirt and Debris at Bearing

1. Deterioration – while deterioration is based on the
    material composition of the bearing and the adverse
    effects of such things as moisture or corrosive
    environment, it is in most instances visible in the
    form of rust or corrosion and is the major cause of
    "frozen" joints or connections. For elastomeric
    material, deterioration is noted by cracks, splits,
    or tears due to excessive stress.

2. Dirt and Debris – this is the prime contributor and
    secondary cause of deterioration. It has a tendency
    to absorb and hold moisture or corrosive material in
    close proximity to the bearing surfaces. In the case
    of rockers, or slotted plates, it may become wedged
    between the bearing surfaces and restrict movement
    of the bearing itself.

3. Bearing Surfaces – inspection of the bearing must
    include a determination that the full bearing surfaces
    are in contact. For example, in a girder span where
    only a partial bearing exists, the load transmitted
    to the support, acting on a partial bearing area, may
    have a bearing value greater than that of the
    concrete support and result in crushing of the support
    or possibly buckling of the steel web over the bearing
    point.

4. Hangers, Windlocks and Keys (Figure 5) - while not
    bearings in themselves, they assist in maintaining
    the alignment of the span while permitting both
    rotation and longitudinal movement. Lateral shear
    keys are particularly important where the bridge is
    skewed or located on a curve.

Figure 5:   Hanger, Windlock and Key

Perhaps the easiest way to cover trusses is to recognize
that there are a great variety of geometrical truss patterns
and to understand that the location and directional orientation
of the individual truss members gives some indication to the
bridge inspector of the type of stress that is normally found
in that member. It is also very important for the inspector
to be able to differentiate between those members which are
considered primary and those that are secondary. A large
percentage of trusses that need to be inspected are in rural
areas. Most of these will either be a through truss, a pony
truss (Figure 6), or a possible combination.

Figure 6: Pony Truss in Rural Areas

One of the most important areas for inspection are the
pins in pin connected trusses where the loads and reactions
on the truss are assumed to act. Should anyone of these members
fail, it would almost certainly be catastrophic. Inspection at
the pin must include an estimate of the amount of wear on the
pin. In some instances, spreaders or washers are used to
maintain the proper spacing of the members on the pin.

There are very few through steel bridges in existence
today, which, at one location or another do not show signs of
vehicular impact (Figure 7). It is usually apparent in the
form of a dent or sharp bend and should be inspected for
evidence of cracks, breaks or tears, damage to connections and
the extent of distortion in the member itself. Significant
damage to a primary member could bring the bridge down as
evidenced by the collapse of the Siloam Bridge in North Carolina
in 1975, when the end post was hit by a car. On the other
hand, if the deformation is in the form of a smooth bow, or an
"S" curve with the member showing equal displacement on both
sides, buckling should be suspect. Buckling due to overstress
is usually indicative of more deepseated problems.

Figure 7:   Vehicular Impact with Endpost

When we think of railings (Figure 8), there are at least four existing hazardous conditions which have been identified from studies of accident information involving railings and vehicles.   These four conditions are:

1. Vehicles penetration of the bridge at approach railing.

2. Snagging of a vehicle by components of the bridge or approach railings.

3. Vehicle collision with the approach end of the bridge or approach railings.

4. Collisions in which a vehicle is redirected by a railing system.

Figure 8:   Example of Railing Deficiency

While old trusses generally have inadequate railing,
solutions have been found to handle this as safely as possible
without infringing too much on the travelway and keeping dead
load to a minimum.

As we move from the inspection of trusses to the inspection
of steel beam and girder bridges, the same problems such as
corrosion and fracture damage are still apparent. In addition,
as members become thinner and deeper, web crippling, web buckling
flange buckling and tension cracks occur. However, of immediate
concern is fatigue cracking.

While it has been recognized that repeated loading may
cause cracking and, ultimately, fracture of structural members,
it wasn't until recently that bridge designers have made extensiv
use of ongoing research in this area. Fatigue cracking under
cyclic loading occurs in two stages - first as a crack initiation
and then as a progressive crack extension (crack growth).

Tests have shown that the total range of stress is a most
important factor in crack growth. The conditions which encourage
fatigue cracking usually are found where the local stress is
high and stress risers may exist, such as at connections,
reentrant or coped sections, indiscriminate welding locations,
and abrupt changes in cross sections (Figure 9).

Figure 9:   Result of Fatigue Cracking

Experience has been that almost without exception, cracks in riveted truss tension members have initiated in the first line of rivets on the outer edge of the rivet group (Figure 10). The end rivet line transmits more than the average load, causing it to be the first location to exhibit fatigue distress. Cracks at rivet holes generally have been found to initiate at the sides of the rivet holes, and to run transverse to the direction of stress.

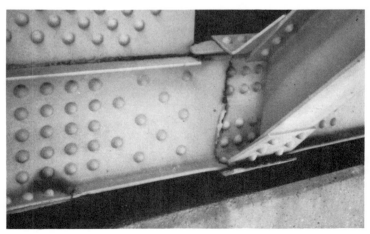

Figure 10:   Cracking at a Riveted Connection

The significant areas in bracing members and systems are similar for trusses and to some extent, girders and beams. Reduction in area, regardless of cause, obviously reduces the capability of the member. In addition, connections to bracing members, gusset plates and primary members are also very important. Riveted, bolted, and welded connections should be inspected for completeness, looseness, signs of movement, and evidence of cracking or buckling in the vicinity of the connection.

The value of visual inspection cannot be overemphasized because an extremely important factor in any inspection is the experience and knowledge required of the inspector. The inspector should be thoroughly familiar with the typical locations, members and details in which cracking may occur. An inexperienced inspector could miss those critical locations. Consequently, the use of a well qualified inspector cannot be overemphasized.

While visual inspection is still the backbone of the inspection program, it goes without saying that the results of research and nondestructive testing have played a significant role in the development of devices (Figure 11) used in bridge inspection and testing. Some of the studies in progress right now may require your expertise and evaluation:

1. Residual Stress Movements in Structural Steels.

2. Development of a Special Eyebar Probe.

3. Detection of Flaws in Reinforcing Steel in Concrete Bridge Members.

4. Acoustic Emission Methods for Flaw Detection in Steel.

5. Determination of Tolerable Flaw Sizes in Full Size Bridge Weldments.

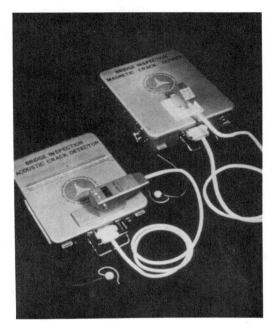

Figure 11:   Nondestructive Testing Equipment
Acoustic Crack Detector and Magnetic
Crack Definer

Your efforts have not gone unnoticed and we will continue to solicit your support in order to assure the traveling public that the condition of our bridges is of major concern.

Thank you.

ASM/ASNT Fourth Annual Forum

14-16 June 1976

PREVENTION OF FAILURE THROUGH

NONDESTRUCTIVE INSPECTION

NONDESTRUCTIVE TESTING OF PRIMARY METALS
— PROCEDURES AND PHILOSOPHY

Keith M. Van Kirk
Magnaflux Corporation

In the late 20's, the burgeoning oil and gas industry was finding it necessary to drill deeper and deeper in the proven and exploratory fields to achieve success. The rotary drilling rigs became bigger and more powerful in order to accommodate these needs. The strings of drill pipe, longer than ever before, were required to carry stresses proportionally higher; and disastrous failures of drill pipe far down in the hole plagued the drillers. The knowledge of fracture mechanics and fatigue was almost nonexistent. However, the examination of the raw ends of the failed pipe did disclose that the preponderance of the catastrophic failures emanated from a crack of some age, and while the failure was erroneously attributed to crystallization because of the crystalline appearance of the final fracture, it was recognized that the prime culprit was an inherent flaw in the material. These flaws were readily apparent by virtue of the discoloration at the point of origin due to the presence of oxides formed during the heating and rolling of the steel. The drillers, logically, asked the steel companies for help in relieving them from these costly and frustrating failures. One steel company, recognizing that the conventional methods of testing for flaws, unaided visual and visual after pickling, were inadequate, approached M.I.T. for assistance in developing more effective methods. Professor A. V. deForest was given the assignment and after considering all the available techniques, experimentally developed and patented the circular magnetization magnetic leakage technique used in conjunction with finely divided magnetic particle sensors for the sure location of all longitudinal and helical surface flaws in the seamless drill pipe. This steel company, although now differently owned, continues to use this technique on these and many other products until this day.

While this is probably by no means the first instance of a steel company accepting responsibility for the surface quality of its products beyond the ordinary requirement, it is typical of the industry's cooperative attitude and of its effectiveness in achieving solutions to its customers' problems.

This spirit of involvement of the steel industry in these, as well as other, critical matters has been a prime factor in the development of the United States into the unchallenged leader of the industrial world. Without high surface quality steels, many of the industrial developments that create this condition of leadership could never have been realized and our high living standards could not have been achieved. Our standards are based on productivity and much of productivity is based on processes that automatically form steel into finished products at high speeds. This highly productive machinery loses much of its advantage if the output is crippled by either down time or expensive hand sorting of its output products resulting from poor surface quality steel. I refer to automatic screw machines, cold headers, bolt makers, spring winding machines, and their like, machines that make critical parts out of long hot rolled or cold drawn bars, straight or coiled, at high rates for long uninterrupted production runs. The oil and gas industries with 20,000 foot plus wells, high pressure pipe lines and torturous refining environments continue to need near perfect steel products to operate with the required dependability and efficiency that keeps us competitive in the world's markets.

The generation of electrical energy by nuclear or fossil fueled power plants requires ever increasing quality to meet operational and environmental demands.

The public, private and military transportation industry blithely continues to design nearer and nearer to the theoretical limits of structural materials — weight and speed being the prime considerations. In fact, in the most critical applications of structures where failure under imposed loads is truly catastrophic as in the operation of missiles and aircraft, a main design criteria is the reliability and capability of those test methods to ensure the absence of minimum sized flaws in fracture critical components.

We can all think of other examples of similar industrial demands which put an ever increasing burden on the steel industry to furnish more and more near perfect material at, hopefully, no inordinate increase in price.

The steel industry not only has reacted to satisfy these demands, but has anticipated them to an overwhelming extent. The steel industry, through the offices of its participant technical societies and related industry technical groups, has been a highly active participant in setting up the guidelines for reasonable attainable quality standard. Working within these guidelines the industry has encouraged the development of N.D.T. methods, techniques, and hardware by independent testing companies and even more so by its own research and development groups. No major steel company in the United States is without multimillion dollar R & D facilities largely devoted to the interest of quality. A very large percentage of each of these facilities is devoted to the development and refinement of nondestructive testing methods. These laboratories have been almost entirely responsible for the development of electromagnetic testing methods as applied to ferrous products. They have made major contributions to ultrasonic testing. They have encouraged the very beginning of magnetic particle testing and they have been responsible for a major breakthrough in the use of penetrating radiation in real-time imaging of critical areas of steel products. Not only have they developed the basic N.D.T. methods to high degrees of sophistication, but they have designed and refined the associated hardware which makes these tests practical quality assurance tools.

United States industry and world industry is in the quiet debt of the steel industry of the United States for its immense but unassuming contributions to technical progress. Indeed, the nondestructive testing industry would not exist were it not for steel's generous contributions.

Even if two steel mills make the same products to the same specifications, no two steel mills are alike. By the same token, neither are the methods by which they achieve their desired or required quality level. It is tempting to hypothesize an ideal mill, from the quality standpoint, for the production of a specific product. However, it is not too difficult for the author to resist this temptation. I am humble before many of far greater knowledge who have trod this path with limited success and I am aware of the many compromises that must be made to achieve any level of harmony in such a devastatingly complex matter as building a mill. However, we can examine a few of the testing alternatives that can be considered.

Regardless of finished product, whether it be forging billets, hot rolled bars, rods, seamless pipe, ERW pipe, SRW

pipe, CW pipe, cold drawn bars, slabs or plates, if quality
is to be assured some surveillance of finished product is
required. Generally speaking, when a product is finished
there is more surface to be tested than in the semifinished
product from which it is made. For instance, a 10 inch tube
round before piercing and rolling has x square feet of surface; after it becomes a tube it has 8x square feet of surface and, assuming a flawless tube round, make a flawless
tube, eight times as much testing must be done to achieve the
same end. Much the same holds true for hot rolled bar products. A 6" x 6" billet has 288 square inches of surface area,
when rolled to one inch bar it is nearly fifty feet long and
has an area of 1700 inches of surface area. If a good billet
makes a good bar, the economy is obviously in the examination
of the billet; further, if the billet is faulty it can be repaired and full value will be retained by removing the flaw,
whereas if the resulting bar is faulty the odds are it will
be scrapped or downgraded or otherwise disposed of in a fashion unfavorable to the economics of the process. Worse yet,
if the bar or rod rolled from the faulty billet is coiled,
examination of the finished product for objectionable flaws
may be impossible. Ideally, many finished mill products
should be made largely flaw free by control of the surface
quality of semifinished steel. Finished product inspection
should be used to detect out of control processes and should
seldom otherwise be needed to sort the good from the bad.
While it is true that some products must be tested in final
form to meet purchase specifications, these requirements do
not obviate the economy of testing and conditioning of the
semifinished products from which they are made. The economics of a high percentage of acceptable finished products are
obvious. The higher the yield, the more able the mill is to
control costs and hold the price line and the earlier the
quality of the finished product can be assured the better the
economics. In any event, whether testing is done of either
semifinished or finished products or both, it is essential
that the test be conducted in an effective fashion at the lowest cost. These two parameters, in combination, have resulted in much research and many developments in the quest of the
ideal solution. As in most crusades of this nature, the
grail of perfection eludes the seekers but nonetheless re-
wards of great value are realized.

Tremendous sums have been profitably expended in developing systems for these purposes and the subsequent expenditure to implement the development often runs into the millions
of dollars for a single testing installation. In the interest

of acquainting you with the Herculean accomplishments of basic steel in the interest of progress, I would like to show you a few illustrations representative of the size and types of these facilities for the testing of some steel products for your benefit.

There is no frivolity associated with product testing in the steel industry, each dollar spent in these pursuits must be fully justified. The process of justification is necessarily extensive and thorough because decisions, once made, involve very large sums of money and many months of planning and preparation. The resulting physical facilities which if unsuccessful are embarrassing reminders of the mistake of those involved.

Most N.D.T. is associated with specified exacting requirement for the finished product. Consequently, specified high quality rod, bar and tubular products get the lion's share of attention. All of these products have as progenitors semifinished steel billets produced by conventional steelmaking practice or by the more recently developed continuous casting process. This semifinished product, almost without exception, must be tested for surface flaws and when disclosed, provision of their removal is mandatory if they are not to reappear in the finished product. Figures 1 through 6 indicate the various, in use, developments for testing semifinished steels. The sum of all these systems in operation today in

Fig. 1. Fluorescent Magnetic Particle Billet Testing Equipment designed for handling a wide variety of lengths and cross sections in an in-line fashion at a rate of about 45 billets per hour.

Fig. 2. Fluorescent Magnetic Particle Billet Testing Equipment designed for handling a limited size range in a high tonnage mill at a rate of about 180 billets per hour.

Fig. 3. Fully automatic Electromagnetic Billet Testing and Marking Equipment designed to handle billets of controlled geometry at 25 to 40 feet per minute, featuring rotating eddy current sensors.

Fig. 4. Fully Automatic High Speed Electromagnetic Billet
Testing and Marking Equipment designed to handle billets of
controlled geometry at a rate of 60 feet per minute, featuring high speed scanning and ultra-sophisticated electronics.

Fig. 5. Fully Automatic Electromagnetic Testing Equipment for
detecting corner flaws in cast and rolled billets. This device automatically programs a four-wheel grinding unit to remove corners containing flaws.

Fig. 6. Fully automatic Ultrasonic Testing Equipment designed to locate internal flaws in steel
billets in two planes and to mark the locations
for removal. This system tests billets at about
100 feet per minute.

the world's steel mills would exceed 100, representing an in-
vestment of at least a quarter of a billion dollars.

The next most frequently tested class of product is tubular goods. Nearly all seamless pipe is thoroughly tested,
as is submerged arc welded line pipe. Electric resistance
welded line pipe and casing also get a thorough treatment, as
do other special-service grades. Testing is done by radiography, real-time and conventional, by magnetic particle, by
electromagnetic, by automatic magnetic leakage field detection
systems and by ultrasound. The variety of combinations on any
one installation can be very impressive. Two installations
are shown in Fig. 7 and 8.

Figure 7 is typical of an electromagnetic testing installation for large-diameter electric resistance welded line

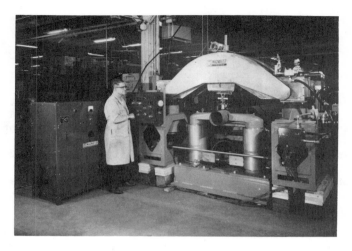

Fig. 7. Electromagnetic testing installation for large-diameter electric resistance welded line pipe.

Fig. 8. Installation of an automatic flaw-detection system using rotating magnetic leakage field detectors.

pipe. The photo shows the instrumentation, sensing head and large magnetizing yokes used to render the magnetic material virtually nonmagnetic through a saturation technique.

Figure 8 is an illustration of an installation of an automatic flaw-detection system using rotating magnetic leakage field detectors. This system is used for testing casing and tubing, both seamless and electric, resistance welded in the full range of diameters and wall thicknesses. It can detect I.D. and O.D. flaws and can differentiate between them. It operates with as high a throughput as 150 feet per minute.

Tubular products are not the only finished products that are thoroughly mill tested. Special surface quality hot rolled bars also get extensive attention with special emphasis on bars to be cold drawn or forged. Because of the large number of bars involved, it is an economic necessity that the testing be done by fully automatic equipment. Figures 9, 10 and 11 indicate the most popular approaches to this type of testing.

Figure 9 shows a typical bar classifier. It is an electromagnetic device using six test stations with nonrotating test fixtures. The bars are driven helically past these stations. When seams are detected, the bars are marked by a high speed cutter on the defect location. These devices process bars at a rate of over 60 feet per minute and are capable of detecting seams as shallow as 0.010 inches deep.

Fig. 9. A typical bar classifier.

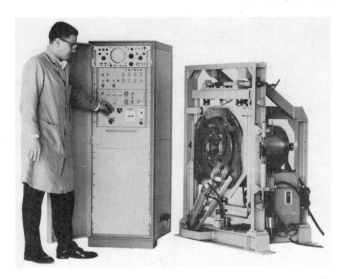

Fig. 10. Device for testing hot rolled steel bars from 1 to 4 inches in diameter at a rate of about 150 feet per minute.

Fig. 11. An automatic fluorescent magnetic particle testing unit for hot rolled bars.

Figure 10 shows a device recently developed by a major steel company's laboratory for testing hot rolled steel bars from 1 inch to 4 inch diameters at a rate of about 150 feet per minute. It features an eddy current sending head that rotates around the bar at up to 1600 RPM. It is also capable of detecting seams as shallow as 0.010 inches and has a marking system which paints the seamy bars with an instantaneously drying paint on the quadrant in which the seam is detected. It is currently in mill operation.

Figure 11 is an illustration of an automatic fluorescent magnetic particle testing unit for hot rolled bars. It features an oscillating high intensity cadmium-helium laser beam for exciting the fluorescent dye and photo detectors to sense the emitted light. The activity of the laser and the detector are coordinated, and the signals are processed by a small computer which can be programmed for pattern recognition, counting, etc. This system serves where the detection of seams less than 0.010 inches is required and is capable of locating seams as shallow as 0.002 inches at a rate of about 150 feet per minute.

These examples of tests of semifinished mill products by no means cover all of the steel industry's efforts to apply nondestructive testing techniques toward the solution of existing and anticipated steel-associated industrial problems.

We have necessarily given very light treatment to the entire subject and have particularly neglected ultrasonic and penetrating radiation — two testing methods with extensive mill application but somewhat less related to the theme of this session. We have also ignored hot product testing.

We have attempted to acquaint those of you who are not already informed with the great depth of involvement of the steel industry in the subject of prevention of premature failure. The steel industry not only solves many existing problems but, by its anticipatory posture in many instances, precludes their very existence.

EFFECTIVE APPLICATIONS OF NDT DURING
PRODUCTION OF AEROSPACE STRUCTURES

J. A. Moore
Vought Corporation
Dallas, Texas

In the production of major industrial items involving
several thousand persons with many different skills, the
quality of the final product is dependent on the efforts of
all parties. A high quality product results from effective
designs, closely-controlled fabrication and assembly operations, and appropriate inspection steps. Due to system
complexities and a large number of detailed operations, most
major industries charge their Quality Assurance Departments
with prime responsibility for the assurance of a high quality
product. To meet this responsibility, the Quality departments perform such functions as design review for quality
allowances, material and tool inspections, monitoring of
fabrication processes, and assembly inspections. Performance
of these functions requires numerous test methods, including
those which are classified as nondestructive tests. In the
past few years, these nondestructive testing (NDT) methods
have become the major tools for quality control during
manufacture of complex structures.

The applicability of NDT in a quality control effort has
been discussed by several authors. McFaul (1) examined NDT
as an adjunct to the design process and product analysis.
Clawson (2) discussed quality control applications of NDT in
the airframe industries which will effectively lower production costs. Frieling and Griffith (3) described in-process
NDT as a tool for quality control. Farnsworth (4) discussed
the relationship of NDT to quality control. Mielnik (5)
illustrated the interrelationship of quality control, reliability, and NDT. The ideas presented by these writers
provide an overview of the part NDT can take in quality
control programs in a variety of industries.

Ensuring the effectiveness of a quality assurance program is to a large degree a management responsibility. Quality assurance personnel must be involved in almost every production step, to an extent necessary to permit no uncontrolled fabrication, assembly, or testing operations. Therefore, Quality management must work closely with the management of other departments to see that quality control, including appropriate NDT, be consistently employed. One result will be the use of NDT in situations where its applicability is well known and recognized. Although the management effort and the routine applications are important and could be extensively discussed, such discussion is not the intent of this paper. Rather, the objective is to discuss NDT methods and applications, those already established and those needed soon in the production cycle, which need refinement or development by NDT engineers.

The discussion is limited to the effectiveness of NDT applications during the design, production, and testing of airframes for aerospace vehicles. A summary of the various applications is presented. With this background, factors affecting the NDT program are examined. The factors include problems which have confronted NDT engineers for some time, as well as new developments such as tighter specifications, new materials and structures, and new or improved NDT techniques. Several areas are defined in which the effectiveness of NDT applications is frequently inadequate.

WHY AN EFFECTIVE NDT PROGRAM

Successful airframe manufacturing, as a business venture in a free-enterprise economy, is based on a continuing program of providing acceptable products to their users, either military or commercial. To accomplish this, the typical corporation organizes its engineering, manufacturing, quality assurance, and other support units such that they perform all of the various functions needed to produce an acceptable product. Regardless of the function of each unit, its efforts must be oriented toward a common objective. This objective could be summarily described as delivery, at the lowest reasonable cost, of a reliable easily-maintained product which will provide optimum service for its intended lifetime. An effective NDT program can participate significantly in meeting this objective.

Two areas in which an effective NDT program has major impact are the reduction of production costs and increased reliability of the final product. Cost reduction is accomplished by monitoring the quality of materials and by controlling or monitoring fabrication processes. NDT techniques are used to prevent faulty or incorrect material from entering the manufacturing cycle, segregate faulty materials or details at intermediate inspection steps, indicate fabrication processes which are almost or actually out of control, and provide additional information needed for disposition of details subjected to material review action. The results are a reduction in the amount of scrap material, higher percentage of production time applied to acceptable materials, and a streamlined flow of parts through the manufacturing cycle. All of these results serve to reduce production costs, if the NDT program is effectively structured and executed.

The need for high reliability of airframes is easily recognized. Failure of airframes for commercial aircraft can result in simply loss of flight time with an accompanied loss of revenue or the more disastrous loss of lives from the crash of a passenger-laden airliner. Failure of airframes for military aircraft can cause the loss of critical flight time, possible loss of lives, and an interruption in the defense system for our nation. Reliability of an aircraft depends on the reliability of a number of systems, including the airframe. The reliability of a well-designed airframe depends primarily on the quality controls applied during production. NDT is a valuable quality control tool, and its value is enhanced by its potential for 100% inspection of production items, such that the soundness of actual hardware is verified just prior to assembly. The need for this capability is becoming even more critical because of the use of new materials and structures, reduced factors of safety, and increased complexities of design.

In the past, NDT methods such as radiography and liquid penetrant have been steadily, although oftentimes slowly, integrated into manufacturing cycles. In recent years, NDT has flourished with the invention of new methods and significant improvements in established methods. However, these developments in themselves have not been sufficient to provide adequate inspection of new aerospace structures. Instead, effective applications of the NDT methods have come from detailed study of what information is needed about a structure, what information each NDT method can provide, and development of inspection techniques which provide the needed information.

APPLICATIONS OF NDT FOR AEROSPACE STRUCTURES

Many applications of NDT during production of aerospace structures are straightforward and in common use by all airframe producers. Table 1 lists the more common NDT methods and their applications. Obviously, these applications are oriented toward metallic airframes of conventional design. Additions to this list can be made when new materials or structures come into common use or when new NDT methods are routinely applied. Whichever techniques are used, they must be closely controlled during inspection of production items. This control is necessary to assure the inspection results are correct and reproducible.

Table 1. Common Applications of NDT for Aerospace Structures

Method	Application
Liquid Penetrant	Surface flaws in forgings, extrusions, weldments, and some castings
Magnetic Particle	Surface and near-surface flaws in ferro-magnetic materials - forgings, weldments, extrusions
Radiography	Flaws (mostly subsurface) in laminated structures, weldments, castings, and forgings
	Assembly Faults
Eddy Current	Surface flaws and flaws in bolt holes
	Conductivity
	Coating and wall thickness
Ultrasonic	Flaws (mostly subsurface) in laminated structures, plate and sheet stock, forgings, and some castings
	Wall Thickness
Acoustic Impact, Infrared, and Sonic	Delaminations and debonds in laminated structures

Some elements subject to control during performance of
the various inspections are listed in Table 2. Those elements
marked with an asterisk are possible areas of improvement.
Other NDT practitioners may well have differing opinions on
the significance of each suggested area of emphasis. The
viewpoints in the following paragraphs are a result of participation in an NDT program oriented primarily toward
technique developments which are readily transferred to production applications. It is recognized that a great deal of
work has been performed in each emphasized area. Hopefully,
forums such as the one in which we are participating will
help define the most rewarding areas of investigation.

Table 2. Some Elements for Control of NDT Methods

Method	Elements
Liquid Penetrant	Materials and equipment* Cleanliness of part surface Application Emulsification Removal Development Interpretation
Magnetic Particle	Materials and equipment Relatively clean surface Flux lines - orientation and intensity* Solution application Interpretation
Radiography	Materials and equipment Part orientation Exposure technique Film processing* Interpretation*
Eddy Current	Equipment Frequency* Field configurations* Display of desired parameter Interpretation
Ultrasonic	Equipment Transducer characteristics* Return signal processing* Interpretation

Liquid penetrant is an established and relatively easily-performed method. A large number of penetrant systems have been developed and their sensitivities experimentally verified. If the application, emulsification, removal, and development steps are performed according to the manufacturers recommendations, satisfactory results are highly probable. Three areas of possible improvement are standardization of present penetrant systems, development of new penetrant systems for special applications, and development of aids for the inspectors in interpretation of indications, for differentiation between flaw indications and false indications. These advances will most likely come from the manufacturers of penetrant systems, with encouragement from users.

Magnetic particle is also a well-established method. Materials and equipment have been developed to handle ferromagnetic airframe parts of widely-varying sizes and configurations. A basic requirement of the method is that magnetic flux lines be generated at all critical points in the part, with their direction perpendicular to probable flaw orientations. This becomes difficult for airframe parts with typically complex shapes. Although we have devices such as the Hall probe to measure external fields, a technique for passively determining the direction and intensity of internal flux lines is needed. In this manner, we could assure ourselves that all critical points are being magnetized sufficiently.

Along with liquid penetrant and magnetic particle methods, radiography (x-ray and gamma ray) is well developed and radiographic techniques for a wide range of materials, thicknesses, and defect types have been optimized. However, it has been recognized that radiographic films contain information not easily discerned by the unaided eye. Improvements are being pursued through video and computerized image enhancement. Another possible area of improvement is the development of new types of films. Again, many of these advancements are coming from equipment manufacturers.

The eddy current method is still due significant improvements in equipment and test procedures. A large number of eddy current systems are in use, in a variety of applications. The method itself is of a nature that it is common for a particular technique to be emperically developed for each application. Technique development would be aided if the parameters affecting the eddy current coil and the field configuration of each coil design could be easily characterized by the users of the eddy current method. This would

perhaps change to a degree the 'highly emperical nature of
present eddy current technique development. Another area of
effort which should be actively pursued is the use of high
frequency (several Mhz range) and multifrequency eddy currents.
Applications for high frequency eddy currents would be testing
of thin-gage materials and low conductivity materials.

The ultrasonic method is in a definite state of flux.
Many successful ultrasonic techniques are being used and many
remain to be developed. Ultrasonics already is or will probably become the most versatile and useful nondestructive test
method. However, at least two elements of the method need
improvement. We have several methods of evaluating transducers, such as schleiren, beam profiling, and distance
amplitude curves, but we still need procedures for accurately
and consistently specifying, producing, and verifying transducer characteristics which influence the performance of
ultrasonic tests. A second element which should be actively
pursued is advanced processing of the reflected or transmitted
signal. A significant amount of information is present in the
modified wavefront, whether the modification was caused by a
flaw or simply by changes in material properties. The techniques for bringing out this information need to be refined
and the significance of resultant data should be determined.

NEW DEVELOPMENTS AFFECTING NDT PROGRAMS

The previous paragraphs have described several elements
of specific techniques which may be preventing maximum
effectiveness of some NDT applications. In addition to these
elements, the aerospace industry is facing several new developments which are also having a significant impact on NDT programs. These developments are outlined in Table 3.

Table 3. New Developments Affecting NDT Programs

Specifications

    MIL-I-6870 (Proposed Revision) - NDI Program
    MIL-STD-1530 - Structural Integrity
    MIL-A-83444 - Damage Tolerance

New Materials/Structures/Manufacturing Methods

    Composites
    Multilaminated Structures
    New Alloys
    New Welding Techniques

218 / J. A. Moore

Table 3. New Developments Affecting NDT Programs

Emerging NDT Methods

    Infrared/Thermal
    Microwave
    Acoustic Emission
    Acoustical Holography
    Optical Holography
    Neutron Radiography

Our customers recognize the importance of an effective NDT program in the production of acceptable vehicles. Consequently, new or revised specifications are being issued to reflect the demand for more effective and reliable NDT. The specifications include MIL-I-6870 (NDI Program Requirements), MIL-STD-1530 (Structural Integrity), and MIL-A-83444 (Damage Tolerance). To meet the requirements of these specifications, airframe manufacturers are obligated to refine their present NDT techniques and, in some instances, develop new techniques.

Another area of rapid development is that of new materials, structures, and manufacturing methods. We are seeing emphasis placed on advanced composites and multilaminated configurations, plus more conventional advances such as new alloys and new welding techniques. These developments are a result of the search for more cost-effective structures with greater performance capabilities. NDT has an active role in the incorporation of these developments into new systems. In many cases, conventional NDT is not adequate for assurance of structural integrity.

As a result of demands such as that just described, the NDT community is experiencing the development of several new NDT methods. These methods, such as infrared/thermal, microwave, acoustic emission, acoustical holography, optical holography, and neutron radiography, have been under development for a number of years (some 10-20 years). Most of the techniques have serious limitations not yet overcome or have applicability under limited circumstances. However, the methods must not be overlooked. Sufficient work must be put into the methods to fully evaluate their applicability and limitations. In turn, the methods should then be appropriately integrated into NDT programs.

SUMMARY OF ITEMS NEEDING ATTENTION

In our quest for an effective NDT program through identification of needed refinements of established techniques and
development of new techniques, we can develop a list of items
which need attention. Such lists have been developed in NMAB-
252 (6) and by an Ad Hoc Group of the Federation of Materials
Societies (7). These references go into greater detail concerning the needs faced by the NDT community. Table 4 might
be a typical list for an airframe manufacturer. Obviously,
each manufacturer must develop its own list, considering the
design of its particular product and any unique inspection
requirements.

Table 4. Suggested List of Items Needing Attention

Flux Intensity and Direction in Ferromagnetic Materials
High Frequency Eddy Current
Eddy Current Field Configurations
Ultrasonic Transducer Evaluation
Advanced Processing of Ultrasonic Response
Acoustic Emission Method
NDT of Multilaminated Structures
Bond Strength Measurement
Surface Contamination Measurement
NDT of Composite Structures
Computerized Data Analysis
Automated Techniques

The first five items in Table 4 were discussed in
previous paragraphs. Of the emerging NDT methods, acoustic
emission is the most promising. As advanced instrumentation
becomes available and NDT practitioners become more skilled
in interpreting acoustic emission data, the method should
approach the status of the five major NDT methods. Although
the requirement that the test part be placed under stress will
limit the techniques versatility, its other advantages should
far outweigh such limitations. Concentrated investigation by
numerous organizations into methods for identifying characteristic signals for the various sources of acoustic emission,
ignoring extraneous signals, and correlating acoustic emission
signatures with physical phenomena will help bring the method
into more widespread use.

Multilaminated structures offer advantages over monolithic structures in areas such as fabrication costs and
resistance to catastrophic failure. However, their future
use will depend heavily on reliable NDT of the structures

following fabrication and during service. The effects of
possible bond and metal flaws on the integrity of multilaminated structures should be established. Subsequently, NDT
techniques for locating and identifying the flaws should be
developed. The next two items in Table 4 go along with this
need. A technique is needed for determining the strength of
bonds. Also, a technique is needed for accurately and rapidly
assuring adherend surfaces are in optimum condition for bonding.

Evaluation and testing of structures made of composite
materials, such as advanced fiber-reinforced plastics, is
progressing rapidly. A large amount of work has already been
performed in establishing NDT techniques for composites. However, the anisotropic nature of the material, with its widely
varying physical properties, causes difficulty in the development and standardization of NDT techniques needed for detecting the many material, processing, and service defects which
can occur. If advanced composites are to become routinely-used
aerospace materials, appropriate NDT techniques must be developed and refined.

The last two items in Table 4, computerized data analysis
and automated NDT techniques are indirectly related to technique refinement and development. The importance of these two
items cannot be overstressed. We must make use of the enhancement of NDT data collection and interpretation offered by
computers, especially the newer minicomputers which have high
capability for relatively low purchase and operating costs.
Of course, computers can also help in automating NDT inspections. Automating the techniques offer several advantages,
not the least of which is the reduction of the human factor.

CONCLUSIONS

The effectiveness of NDT programs depends on a number of
factors, including effective management interaction. Another
major factor is the recognition by all concerned of problems
with established techniques and the need for improved or new
techniques, brought on by developments such as new specifications, materials, structures, or manufacturing methods. A
discussion of several such items has been presented in this
paper. Naturally, the list is not exhaustive. Each manufacturer can add to the list, resulting from his own particular inspection requirements. Whatever the driving forces, a
concentrated effort by all NDT organizations is needed to
provide solutions to these problems.

REFERENCES

(1) McFaul, Howard J., Materials Evaluation, Vol. XXX, No. 4, 1972.
(2) Clawson, B. W., Materials Evaluation, Vol. XI, No. 3, 1953.
(3) Frieling, G. H. and Griffith, E. C., Materials Evaluation, Vol. XXXI, No. 3, 1973.
(4) Farnsworth, William P., Brig. General, Materials Evaluation, Vol. XIII, No. 6, 1955.
(5) Mielnik, E. M., Materials Evaluation, Vol. XXXII, No. 4, 1974.
(6) Nondestructive Testing, NMAB-252, National Academy of Sciences-National Academy of Engineering, Washington, D.C.
(7) Conservation in Materials Utilization, Ad Hoc Task Group on National Materials Policy Study, Federation of Materials Societies, Materials Evaluation, Vol. XXXI, No. 4, 1973.

"Q-T"
QUANTITATIVE TESTING
WITH
INSPECTION PENETRANTS

J. R. Alburger
Shannon-Glow, Inc.
7356 Santa Monica Boulevard
Los Angeles, Calif. 90046

IN BRIEF:

The discussion presented here should be of interest to scientists and engineers who are concerned with the manufacture, specification, and control of inspection penetrant materials, and their ultimate use in applications where precise evaluations of flaw magnitudes are wanted. Instrumentation and techniques of measurement are described.

*

THE INSPECTION PENETRANT PROCESS - ITS PRIMARY FUNCTION

The fluorescent penetrant process, regardless of its mode of usage, is primarily a tool for detecting and locating flaws in fracture-critical parts. If penetrant process materials are selected to provide suitable levels of flaw detection sensitivity, and if the materials are utilized correctly, so as to yield a satisfactory level of flaw-entrapment efficiency, then crack defects of appropriate magnitudes in test objects may be detected and identified as to their locations.

Up until recently, fluorescent inspection penetrant processes have been limited in their capabilities to a qualitative evaluation of defects, to demonstrate that they exist and to show where they are located. Usually, by close examination, it is possible to tell something about the flaw configuration, as for example whether it is a linear or jagged crack, or whether it is a pinhole, or crazing, or any

one of several other types of defects. Also, by considering
the apparent brightness of a flaw indication, it is possible
to tell whether the defect is large or small, but such
evaluations are normally qualitative in nature, since the
conventional penetrant processes are incapable of providing
accurately reproducible results and levels of flaw-entrapment
efficiency which are high enough to yield indication brightness values which are representative of entrapments which fill
the flaws completely.

INDICATION BRIGHTNESS AS A FUNCTION OF FLAW MAGNITUDE

If we consider the Beer's Law fluorescence transition
curve for a given penetrant, we see that the brightness of
fluorescence response is a function of the thickness of a
layer of the penetrant. To put this in the perspective of a
micro-entrapment of penetrant in a flaw, it is seen that the
effective size of the penetrant entrapment has a direct
relationship to the size of the flaw, and it should therefore
be possible to determine the size of a flaw by evaluation of
the flaw indication.

For such an evaluation to be meaningful, at least two
requisites must be satisfied. First, the flaw must be
completely filled with penetrant, at least at some point in
time, so that the evaluation or measurement which is made
will be representative of the flaw *volume*. Second, if the
flaw exists in an opaque (metal) surface, the entrapment
which is formed in the flaw must be drawn out by a suitable
development procedure, so that it can be seen and/or
measured as to its brightness.

NON-QUANTITATIVE FLAW ENTRAPMENTS

In conventional inspection penetrant processes, the
penetrant is first applied to the surface being tested, then
excess surface penetrant is removed for the purpose of
eliminating interference from residual smears of penetrant
and background indications of non-significant porosities.
With conventional penetrants, it is virtually impossible to
remove residual thick films of surface penetrant without also
incurring a substantial depletion of entrapments in actual
flaws.

Flaw-entrapment efficiency is defined as the amount of
penetrant which remains in a flaw (in percent) relative to

the maximum amount of penetrant which can be contained in the
flaw (volume of the flaw). Typical penetrant processes act
to rapidly strip penetrant entrapments out of flaws, either
by over-emulsification in the case of P/E-type penetrants, or
by over-washing in the case of W/W-type penetrants. Thus,
the flaw-entrapment efficiency of a typical fluorescent
penetrant process may often be on the order of only a few
percent. Conventional penetrant processes do not lend
themselves to easy enhancement of flaw-entrapment efficiency
by shortening the contact time of the emulsifier, wash water,
or solvent remover, as the case may be, since the depletion
of flaw entrapments is so rapid that a very substantial loss
of the entrapments may occur within the first few seconds of
contact with the remover.

EVALUATION OF FLAW MAGNITUDES USING
NON-QUANTITATIVE FLAW ENTRAPMENTS

Some penetrant process materials, notably the P-600 -
Series Diamorphic Penetrants, are characterized by a
relatively slow response to remover (wash-water) action, and
the resulting depletion follows an exponential curve which
corresponds with reasonable accuracy to the theoretical
curve of diffusion depletion. Thus, by using such penetrants,
and calibrating them with respect to dye sensitivity and
depletion time constants, it is possible to work backward
from the relative brightness of a flaw indication which is
obtained after a measured time of remover contact, and derive
the equivalent magnitude of the entrapment (flaw size) which
would have pertained at the outset before depletion started.
Mathematical derivations of this kind have been discussed in
some detail in a previous publication. (1)

"QUANTITATIVE" FLAW ENTRAPMENTS

There are now at least three types of inspection penetrant processes which have been developed for the purpose of
providing high levels of flaw-entrapment efficiency. All of
these make use of certain chemical methods of solubility-
inhibition or inhibition of emulsifier action.

In the case of P/E-type penetrants, low-energy
emulsifiers may be utilized so that the activity of the
emulsifier in stripping out entrapments of penetrant from
flaws is greatly reduced. (2)   More important, though,

has been the development of a unique process material and a
technique for removing thick-film residues of penetrant from
test surfaces without causing any depletion of entrapments
in actual flaws. This is the G-201 Film-Breaker material
and process. (3)

In the case of W/W-type penetrants, these too can be
slowed down with respect to their rate of wash-removal, so
that they dissolve very slowly in wash water, and are therefore less subject to rapid depletion of flaw entrapments.
Materials of this kind are known as the Tracer-Tech Slow-
Solubility Inspection Penetrants. (4)   Here, too, the G-201
Film-Breaker remover may be utilized, (3) so that thick-film
residues of penetrant can be removed from test surfaces
without any material depletion of entrapments in flaws. For
application of the G-201 Film-Breaker Stripper to the Slow-
Solubility penetrant process, an added requirement is that
the Film-Breaker stripper-wash mixture must be inhibited so
that it cannot dissolve the penetrant and strip it out of
actual flaws. This inhibition of solubility is accomplished
by saturating the G-201 stripper mixture with dissolved
penetrant. (5)

In the case of solvent-remover-type penetrants, the
solvent remover may be inhibited in its action, so that it
will not dissolve penetrant entrapments very rapidly. (6)
Here again, in a manner similar to the retarded P/E process
(Spray-Scrubber), thick-film residues of penetrant may be
removed through use of the G-201 Film-Breaker stripper, and
since the penetrant is insoluble in water, entrapments in
surface flaws are not stripped out and lost.

In all cases, where the G-201 (or similar) thick-film
penetrant removal technique is employed, test parts which are
so treated exhibit almost 100% flaw-entrapment efficiency.
This level of entrapment efficiency is too great to be
practical for the normal detection and evaluation of flaws,
since we usually find large numbers of porosity background
indications which tend to interfere with actual crack
indications. However, a high level of flaw-entrapment
efficiency, sufficient to provide good stability of flaw
entrapments during the removal of thick-film residues of
penetrant, is very advantageous where it is desired to
measure or otherwise evaluate flaw magnitudes to a high
degree of accuracy.

QUANTITATIVE EVALUATION OF FLAW MAGNITUDES

Assuming that we can generate a flaw entrapment of penetrant which completely fills the flaw under study, and further, assuming that we can develop the entrapment so as to draw it out of the flaw to a point where it can be measured as to its brightness, then it will be possible to determine the magnitude of the flaw in terms of an equivalent film thickness of a layer of fluorescent penetrant liquid. There are, of course, some shortcomings in this concept of expressing flaw magnitude, since the result obtained in the evaluation has only one dimension (film thickness), while the actual flaw has at least two significant dimensions (depth and width), and sometimes we must take into account the entire volume of the flaw (length, depth, and width). Even with this limitation, we are still able to get a handle on flaw magnitudes, in a manner which permits us to rate flaws relative to one another as to their apparent sizes.

An important consideration in connection with measuring a developed indication as to its relative brightness, is that the developer operation must not enhance the effective brightness of the entrapment. In other words, the brightness of the developed indication should be about the same as the brightness of the original undeveloped entrapment *would have been* if the test surface was completely transparent. Most of the modern developers contain ingredients which enhance fluorescent brightness, (7) or which provide a reflective background (white pigment), thus augmenting the apparent brightness of flaw indications. However, there is one developer, the Tracer-Tech D-499C material, which provides efficient extraction of entrapments from flaws, high resolving power to yield sharp and well-defined indications, and without any substantial shift in the apparent brightness value of entrapments which are developed out of flaws in transparent surfaces (fractured glass).

It should be noted that even if there is no effect of chemical-brightening in the developer action, some brightening will often occur, merely because of the change in entrapment geometry which results from development. For example, a crack defect which is fairly tight may provide an initial entrapment which is quite deep but at the same time appears as an extremely thin liquid film. When such an entrapment is drawn out by a developer, it may become concentrated so that its depth-width ratio may be altered drastically. This alteration in depth-width ratio is often accompanied by a shift in apparent brightness.

In any event, if we establish a standard procedure for development of flaw indications, and then measure or judge the brightness of these developed indications relative to some suitable reference standard, we will obtain quantitative values of brightness which can be translated into equivalent flaw magnitudes.

TWO ESSENTIAL MEASUREMENTS

If a given penetrant process is to be employed in a manner which will enable us to determine the magnitude of a flaw, there are two measurements which must be made. First, the penetrant must be calibrated as to its dye-performance sensitivity. This involves the determination of the alpha (α) value for the penetrant. As for dye-performance sensitivity, all Tracer-Tech inspection penetrants are accurately calibrated and standardized in manufacture, so that the data supplied for a given penetrant can usually be used with confidence in connection with the procedures for evaluating flaw magnitudes which will be described below.

Second, a measurement must be made of the brightness of the flaw entrapment (as developed), relative to the brightness of a thick film of the penetrant. For purposes of evaluating flaw magnitudes, we are not concerned with interference effects of background indications. Such indications can be ignored, since our brightness measurements will focus on the flaw indication to the exclusion of any background effects which might be present. There are exceptions to this, though, such as cases where we wish to evaluate the magnitudes of craze-crack flaws, or porosity features in a test surface, where the flaws are not sharply defined. In such cases, the best we can do is to try to isolate an individual point in the flaw pattern, and measure that particular spot as to its apparent flaw magnitude in terms of its equivalent film thickness of the liquid penetrant.

FLAW SIZES WHICH CAN BE MEASURED

Measurements of flaw magnitudes are accomplished through use of an appropriate fluorescent penetrant which is calibrated with respect to its Beer's Law transition characteristic. A brightness measurement is made for the developed entrapment of penetrant which is drawn from the flaw, and as long as this brightness value falls somewhere on the sloping portion of the

Beer's Law transition curve, a corresponding value for flaw
magnitude can be determined. Any given penetrant will provide
for flaw-magnitude evaluations over a range which corresponds
to the portion of its transition curve from relative brightness
values of about .05 up to about .95.

For most practical purposes, a single penetrant, such as
a Level 7 penetrant, can be used for the determination of
significant flaw magnitudes. The Level 7 penetrant can be
employed for the determination of flaw magnitudes (Tau values)
ranging from about .00001 cm. up to .0005 cm. (.1 micron to
5 microns). This range may be extended through use of other
penetrants having greater or less dye-performance sensitivity.

CALIBRATION CURVES FOR PENETRANTS

Figure 1 shows two calibration curves for fluorescent
penetrants of Tracer-Tech Levels 1 and 7, respectively. The
Level 7 sensitivity is commonly available in commercially-used
fluorescent penetrants. The Level 1 is supplied as a
specially calibrated penetrant intended for laboratory
analysis of flaw magnitudes.

These two sensitivity levels provide a range of flaw-
magnitude measurement capability from about .00001 cm. up to
.004 cm. (.1 to 40 microns). Any high-stability P/E-type
penetrant may be employed for measuring flaw magnitudes, but
the penetrant must first be calibrated by the procedures which
are outlined below, so as to determine its alpha (α) value.
Once this is done, it is a simple matter to draw the
appropriate fluorescence transition curve on the chart of
Figure 1.

Since the calibration curves illustrated in Figure 1 are
(theoretically) simple exponential curves, and always have the
same shape regardless of their position on the chart, any
particular curve may be located on the chart merely by making
one measurement. Usually, it is most convenient to make a
measurement which corresponds to the mid-point, or inflection
point, of the transition curve. For a fluorescence transition
curve, we may measure the film thickness at which the fluorescent brightness is .5 relative to the maximum (thick-film)
brightness.

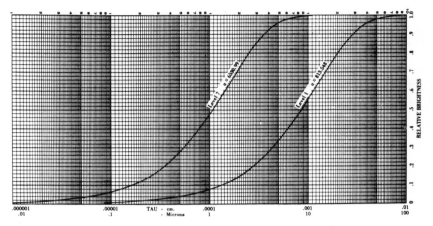

FIGURE 1 - Tau Values Plotted as a Function of Relative Fluorescent Brightness

INSTRUMENTS AND APPARATUS

Various instruments which are suitable for measuring penetrant performance parameters have been described previously, (8) and detailed procedures for calibrating penetrant materials, test panels, and reference plaques, have been set forth in another publication. (1) Similar instruments and procedures may be utilized in connection with the calibration of the Q-T penetrants and the evaluation of flaw magnitudes in test objects. The following items are required for making measurements of the two basic performance features which are of interest here.

1. **A Darkened Black Light Booth** - All fluorescence measurements should be carried out under conditions of subdued white light in a darkened laboratory area. A small amount of white light is permissible so that micrometer and meter readings may be made, however ambient white light intensity should be kept as low as possible. A hooded booth or work table equipped with black cloth drapes is adequate.

2. **A Black Light Lamp** - This should be a 100 watt mercury arc spot lamp (Shannon-Glow Model L-121 Spot). The lamp should be powered by a voltage-regulated transformer (Sola-type), or a phase-controlled voltage-regulating dimmer (Shannon-Glow). The black light lamp must be mounted on a swivel bracket in such a way that it may be directed downward at an angle of about 45° to irradiate the stage of a micro-

scope. A knob or handle should be provided on the back of
the lamp to permit manual positioning and direction of the
black light beam. A heavy-duty rheostat may be inserted in
the primary circuit of the black light transformer, so as to
permit fine adjustment of the black light intensity. If a
sufficiently sensitive photomultiplier detector is available,
a 4 watt tubular-type black light lamp may be used in place
of the mercury arc lamp.

3. A Micrometer Stage Microscope - A monocular microscope is required which has a 10x Huygens eyepiece. The
stage of the microscope shall be movable (on ball bearings),
and shall be controlled by screw-type micrometers which are
calibrated in mm., with an accuracy of reading down to .001
mm. (1 micron). The main traverse of the micrometer stage
shall be approximately at right angles to the direction of
the black light beam. The micrometer which controls the
main traverse shall have a linkage (through a flexible shaft
or equal) to a drive motor. A removable bracket shall be
provided on the microscope stage to serve as a positioning
stop for test parts. A 10x objective lens should be used.

4. A Stage-Drive Motor - A stepping motor or similar
drive motor is required for scanning over a test specimen
which is placed on the movable microscope stage. Preferred
equipment for this purpose is the stepping motor and pulse
generator apparatus supplied by Schlumberger (Heath-Malmstedt-
Enke). Connect this drive motor through a flexible shaft to
the main-traverse micrometer screw on the movable stage of
the microscope. Synchronize this motor with the strip-chart
recorder drive.

5. A Strip-Chart Recorder - This item is preferably
the type supplied by Schlumberger (Heath). This equipment is
preferred, since the chart drive is by means of a stepping
motor, and this can be synchronized with the drive motor for
the mechanical stage of the microscope. Chart speed should
be adjusted so that an even number of divisions (10 for
example) equal 1 mm. of the micrometer movement. This is
necessary so that spot diameters may be conveniently measured
to an accuracy of .01 mm. or better. Voltage sensitivity of
the recorder should be 1 volt full scale. Slew rate should
be rapid enough so that accurate response is provided in
scanning over narrow crack indications.

6. A Photomultiplier Detector - A high-sensitivity
photomultiplier photocell is required, preferably an eleven-
stage phototube. The photocell must be mounted on a 10x

Huygens eyepiece having an 80 micron pinhole in its reticle
plane. In this manner, the microscope may be focused on a
fluorescent field using a full-view Huygens eyepiece, whereby
an image of the fluorescent field appears in the eyepiece
reticle plane. Then, when the photocell assembly is inserted
into the microscope eyepiece tube, the photocell sees an area
in the fluorescent field which is about 8 microns in diameter.
The photocell and eyepiece lens assembly should be equipped
with a set of optical filters which serve to cut out ultraviolet radiation and correct the photocell response to
approximate the spectral response of the human eye.

7. <u>Photocell Battery Pack</u> - Batteries should be used
to supply high voltage to the photomultiplier detector. This
is desirable so as to minimize meter fluctuation and jitter.
Some jitter is unavoidable, due to the 60 cycle input to the
black light lamp, and line fluctuations which cannot be
smoothed out by the voltage-regulating transformer. The
battery pack should permit a step-wise adjustment of voltage
from zero to 1500 volts, depending on the voltage limit of
the phototube.

8. <u>A Photocell Meter</u> - The photocell meter should have
a large taut-band meter, preferably with a 4-1/2" scale. It
should have provisions for zero-set to balance out dark
current, and a sensitivity control to permit full-scale
meter readings for standard reference plaques or reference
thick-film layers of penetrant. A connector receptacle
should be provided for connection to the strip-chart recorder,
and a screw-set adjustment should be provided which permits
adjustment of the output voltage to the point where full-scale
on the meter corresponds to 1 volt at the meter output
connector terminals. The photocell, the battery pack, and
the meter, as described, are all available, as a package from
Tracer-Tech, as Meter Model Y-111.

9. <u>A Meniscus-Method Large-Lens Set</u> - This is the
Tracer-Tech Model Y-202 Outfit. (9) This item is required
for spot diameter measurements in the determination of dye-
performance sensitivity. The 3-1/2" diameter glass lens
(Y-202D) must be ground to a surface curvature of .5 diopter
(radius of curvature is 106 cm.). A brass ring weight is
mounted on the glass lens to keep the lens from floating on
liquid films during measurement tests. This set included a
flat optically-polished black glass platen (Y-202A) for use
with fluorescent penetrants. The set also includes a clear
glass platen (Y-202B), an opal-glass diffuser plate, and two
color-contrast filters.

10. <u>An Electronic Calculator</u> - For the purpose of data
reduction and computations, a computer or electronic calculator
is required which will handle exponential functions and natural
logarithms. The Hewlett Packard HP-45 calculator is suitable
for this purpose. Tables of logarithms may be utilized, if
desired.

11. <u>Miscellaneous Apparatus</u> - Some miscellaneous items
are needed as follows: A 2 cm. spacer block for off-setting
the movable microscope stage, Flat-tipped spoonulas or similar
applicators, Methylene chloride and dimethyl formamide in
plastic squeeze bottles with flip-up spouts, for cleaning
penetrant from test parts.

<u>MEASUREMENT METHODS</u>

All measurements are made in terms of fluorescent
brightness response under standardized conditions of ultraviolet irradiation of the test specimen. In most cases, we
are interested in brightness *ratios* with respect to a
reference material, rather than absolute values of brightness.
Thus, the reference material is set up so that the photocell-
meter reads 100% full scale. Then, the test specimen is
positioned so that it receives the same irradiation as the
reference material, and the test specimen brightness reading
is stated as a decimal value of the full-scale reading
(from 0.00 to 1.00). The reference meter setting need not be
exactly full-scale, in which case the proportional meter
reading is calculated.

<u>DYE PERFORMANCE SENSITIVITY</u>

This measurement involves the determination of the
distance between points of half-brightness across the circular
nonfluorescent black spot which appears in the Meniscus-Method
setup. (9)

When a meniscus-shaped layer of fluorescent liquid forms
between the lens and platen in the Meniscus-Method setup,
fluorescence response approaches zero at the point of contact
between lens and platen. At distances exceeding 2 cm. from
this contact point, the fluorescence response levels off at
a maximum (for typical fluorescent penetrants), and this
brightness level may be used as a reference (100% on the
meter). The puddle of fluorescent liquid must be at least
2" in diameter, in order to avoid errors due to edge effects

and internal reflections of light within the lens. Also, scanning by the microscope-photocell must be in a direction at right angles to the black light beam, so as to minimize errors due to internal reflections within the lens.

Step 1 - Place the black glass platen on the movable microscope stage. Apply a puddle of test penetrant on the center of the platen, sufficient to provide a circular patch of penetrant under the curved lens at least 2" in diameter. Place the ring-weighted glass lens on the puddle, and rotate the lens gently to insure the formation of a clean black spot at the point of contact between lens and platen. After positioning the lens on the platen, use care when moving the microscope stage to avoid shifting the lens position.

Step 2 - Insert the full-view eyepiece in the microscope tube, and adjust the microscope focus. (The microscope should have a 10x objective lens). Push the movable stage over to a point where the edge of the fluorescent puddle can be seen in the eyepiece, and use this edge for focus. After the lens and platen have been in use for some time, microscopic scratches can be seen at the center of the black spot, and these may be used to focus the microscope.

Step 3 - Set the main-traverse micrometer screw so that it permits free movement of at least 5 mm. in either direction. Carefully slide the platen back and forth on the movable stage, until the black spot is exactly centered in the eyepiece field.

Step 4 - Remove the full-view eyepiece, and plug the photomultiplier photocell into the microscope eyepiece tube. Connect the photocell to its battery pack and the meter. Connect the strip-chart recorder to the meter.

Step 5 - Push the movable stage off-center a distance of 2 cm., using a 2 cm. spacer block between the micrometer screw and the stage anvil. Adjust the black light and/or the photocell voltage and/or the meter sensitivity so that the meter reads exactly 100% or 1.00 (full scale).

Step 6 - Place an opaque card so as to interrupt the black light beam, and adjust the meter to read zero. Re-check the full-scale setting.

Step 7 - Remove the 2 cm. spacer, being careful not to disturb the placement of the lens on the platen. Back off the main-traverse micrometer screw until a meter reading greater than 50% of full scale is obtained.

Step 8 - Start the recorder and stage-drive motors. Scan across the black spot until the meter reading again exceeds 50% of full scale.

Step 9 - Measure the distance (number of divisions) between points on the strip-chart trace corresponding to 50% of full-scale deflection. Convert the measurement to mm. or cm. This reading is the *Spot Diameter* of the test penetrant.

DATA REDUCTION IN MENISCUS-METHOD MEASUREMENTS

Film thickness (t), at inflection, is derived from the lens geometry:

$$t = \frac{r^2}{2R} = \frac{r^2}{212} \tag{1}$$

Where:
r = Radial distance from spot center (cm.)
D = Spot diameter (cm.)
R = Radius of curvature of lens surface (106 cm.)

This translates to:

$$t = .004717 \left(\frac{D}{2}\right)^2 \tag{2}$$

For example, Level 7 penetrants exhibit a spot diameter of .3 cm. (3.0 mm.). The t value for this spot diameter is .000106 cm. (See dye-sensitivity curve of Figure 1 at its inflection point).

Alpha (α) values are derived from Beer's Law (Bouguer's Law):

$$\text{Relative Brightness} = .5 = e^{-\alpha t} \tag{3}$$

$$\log .5 = -\alpha t = -\alpha .004717 \frac{D^2}{2} \tag{4}$$

$$\alpha = \frac{-\log .5 \times 4}{.004717 \ D^2} = \frac{587.786}{D^2} \tag{5}$$

SUMMARY OF MENISCUS-METHOD MEASUREMENTS

Measurements by the Meniscus Method yield data in the
form of spot diameters for fluorescent liquids. The spot
diameter values may be converted into α values (absorption
coefficients). Conversion to α values must be made if the
penetrant calibrations are to be utilized for determination
of flaw magnitudes.

MEASUREMENT OF RELATIVE BRIGHTNESS OF A FLAW ENTRAPMENT

This procedure may involve measurements on actual
service parts, and since such parts come in a variety of
shapes and sizes, the measurement method must be adapted to
the part being tested and the location of the flaw on the
surface of the part. In outlining the following procedure,
it is assumed that a flaw has been located, and its location
marked on the test part by circling it with a wax pencil.

Small test parts may be examined and measured by mounting
them on the stage of a microscope which is adapted to use the
photomultiplier photocell. For large parts, we may have to
bring the microscope/photocell and black light lamp to the
part, rather than vice versa. In any event, the microscope/
photocell apparatus must be arranged, by means of suitable
clamps and brackets, so as to focus on the flaw indication
being studied. Also, the black light lamp must be clamped in
a position such that it irradiates the flaw and its developed
entrapment. Allow the lamp ample room to swivel, so that the
intensity of the black light beam on the flaw area may be
adjusted.

Step 1 - Clean the test surface thoroughly, and treat
it with a high-stability penetrant (P-147 or P-141). Allow
a dwell time of 5 to 10 minutes.

Step 2 - Condition the test surface by spraying it
thoroughly with water containing the G-201 Film-Breaker
Stripper. In many cases a strong spray of warm water (at
110° F.) will suffice. Dry the test surface with compressed
air. At this point, the test surface should show considerable
background fluorescence, but when the surface is touched or
wiped with a cleansing tissue no fluorescence should be
picked up or marked off on the tissue.

Step 3 - Apply the D-499C Developer, sparingly, to the
flaw area, and allow a development time of 10 minutes.

Develop fully, so that further application of the D-499C does
not produce a further increase in apparent brightness. For
cracks in transparent surfaces (glass or anodic coatings), no
development is required.

Step 4 - Position the test part under the microscope,
using appropriate blocks and stops, so as to focus on the flaw
being evaluated. For large test objects, arrange the microscope on a sturdy bracket so that it is focused on the flaw.
In either case, a rack and pinion micrometer movement, a
swivel bracket, or a movable stage, should be utilized so
that the brightest part of the flaw indication can be brought
under the microscope.

Step 5 - Insert the full-view eyepiece in the microscope tube, and focus on the flaw indication. Center the
flaw indication in the eyepiece field.

Step 6 - Plug the photomultiplier photocell into the
microscope eyepiece tube, connect it to the meter, and
position the black light lame so that a meter reading is
obtained. Move the microscope or the movable stage so as to
obtain a maximum reading on the meter.

Step 7 - Prepare a strip of blackened brass shim stock
with a drop of the penetrant close to one end. While holding
the penetrant drop in the microscope field (on the test
surface), being careful not to disturb the flaw indication,
adjust the position and direction of the black light lamp so
as to produce a full-scale reading on the meter. This value
is B_o.

Step 8 - Remove the shim strip with the penetrant
reference drop, and take a meter reading for the flaw
indication. This value (as a decimal value relative to
full-scale) is B_1.

Step 9 - If the relative brightness of the developed
flaw indication (B_1/B_o) is greater than .9, repeat the test
using a lower-sensitivity penetrant. If B_1/B_o is less than
.1, repeat the test using a higher-sensitivity penetrant.
When a satisfactory value of relative brightness is obtained
(between .1 and .9), record the value and proceed with data
reduction.

The Level 7 of the P-147 material is the highest level
of dye-performance sensitivity which is available in the

P-140-Series penetrants, for the reason that the sensitizer
dye concentration is close to saturation at this level.
Special research penetrant formulations can be made having
sensitivities which are several levels higher, but these
must be used in high-temperature environments in order to
maintain the sensitizer dye in solution.

DATA REDUCTION IN FLAW MAGNITUDE MEASUREMENTS

Flaw magnitudes are calculated and expressed in terms
of Tau (τ) values, where Tau is the thickness of a layer of
penetrant which would yield the same relative brightness as
the measured and calculated value for the developed flaw
entrapment. Since the penetrant has been calibrated, and we
know its alpha value, we can determine Tau from:

$$R.B. = B_1/B_0 = 1 - e^{-\alpha\tau} \tag{6}$$

or

$$\tau = - \frac{\log(1 - R.B.)}{\alpha} \tag{7}$$

SUMMARY OF FLAW MAGNITUDE MEASUREMENTS

The above-described method of measurement and data
reduction yields Tau values which are representative of the
dimensional magnitudes of flaws from which measured
indications are derived. Once the penetrant has been
calibrated and its alpha value determined, then the
magnitudes of various flaws may be evaluated, each by making
only one measurement of its relative brightness.

The P-141 and P-147 penetrants are preferred for use in
measuring flaw magnitudes, for the reason that they form flaw
entrapments which are extremely stable in the presence of
water. Other types of penetrants may exhibit a substantial
degree of leaching of the fluorescent sensitizer dye during
the surface-conditioning operation, thereby invalidating the
measurement.

COMPARISON TESTS WITH A REFERENCE SPECIMEN

For many inspection applications, reasonably accurate estimates of flaw magnitude may be made by means of a side-by-side comparison of the developed flaw indication with a calibrated Y-404 Fractured-Glass Test Panel. The Y-404 panel has a series of six graduated steps on its surface which have micro-fractures ranging in size from as small as about .00005 cm. to .002 cm. (.5 micron to 20 microns).

Calibrations for this type panel are carried out as described above, and in use the panel is treated with the same penetrant as is used on the test part, and is surface-conditioned in the same manner, using the G-201 Film-Breaker stripper (or a hot water spray). No development is needed, since the Y-404 panel is relatively transparent.

When the penetrant-treated Y-404 test panel is placed beside the developed flaw indication under study, a value judgement may be quickly and easily made as to which step on the Y-404 panel is closest in brightness to the flaw indication. Reasonably accurate interpolations may be made where the flaw indication brightness falls between steps on the Y-404 panel.

Relative brightness values may often be estimated with good accuracy by making visual value judgements in comparing the developed flaw entrapment with the reference drop of penetrant. Regardless of how the R.B. value is determined, it can be inserted in Equation (7) for calculation of the Tau value. Alternatively, Tau values may be determined graphically by reference to the appropriate Beer's Law calibration curve for the penetrant being utilized (Fig. 1).

Table I may be utilized for a quick determination of the Tau value which corresponds to a given value of Relative Brightness (R.B.). After some experience, an inspector can estimate values of Relative Brightness with reasonable accuracy, without actually measuring the brightness, and then refer to the table for determination of corresponding Tau values. Similar tables may be easily constructed for any high-stability penetrant having a known α value.

SUMMARY

The advent of high-stability calibrated penetrants, and the G-201 Film-Breaker stripper along with its mode of usage in "conditioning" test surfaces, have made it possible to generate flaw entrapments which are close to 100% filled, and which have known Beer's Law transition characteristics. By properly developing the entrapment in a given flaw, and by measuring its brightness relative to a thick film of penetrant, a value (Tau) can be determined which is a measure of the flaw magnitude.

TABLE I - TAU VALUES (Microns)

$$\tau = \log (1 - R.B.) / \alpha$$

R.B.	Level 1	Level 7	R.B.	Level 1	Level 7
.05	.63	.078	.55	9.82	1.22
.10	1.29	.161	.60	11.26	1.40
.15	1.99	.249	.65	12.90	1.61
.20	2.74	.342	.70	14.80	1.84
.25	3.54	.440	.75	17.04	2.25
.30	4.38	.546	.80	19.78	2.46
.35	5.30	.660	.85	23.32	2.91
.40	6.28	.782	.90	28.30	3.53
.45	7.35	.915	.95	36.82	4.59
.50	8.52	1.06			

REFERENCES

1. Alburger, J. R., "Theoretical Aspects of Penetrant Entrapments and Fluorescent Indications". Presented before the Am. Society for Nondestructive Testing. Detroit Convention, October, 1974.
2. U.S. Patent No. 3,422,670
3. U.S. Patent Pending
4. U.S. Patent No. 3,896,664
5. U.S. Patent Granted
6. U.S. Patent Granted
7. U.S. Patent Granted
8. Alburger, J. R., "Instruments and Test Methods as Employed in an Improved Penetrant Material Specification." Presented before the 9th Symposium on Nondestructive Evaluation, April, 1973, San Antonio, Texas.
9. U.S. Patent No. 3,107,298

THE DEVELOPMENT OF QUANTITATIVE NDI FOR RETIREMENT-FOR-CAUSE

John E. Allison
Air Force Flight Dynamics Laboratory
J.M. Hyzak and W.H. Reimann
Air Force Materials Laboratory

ABSTRACT

To investigate the viability of quantifying NDI signals, relationships between eddy current signals and flaw dimensions were examined. The prime emphasis of this paper is exploring these relationships and the difficulties encountered in establishing them. A fracture mechanics/NDI-based disk rejection criteria, retirement-for-cause, which requires quantitative NDI, is also described. The case of radially cracked bolt holes in a turbine disk inspected by a semiautomatic eddy current technique is used to illustrate the procedures.

Monotonically increasing relationships, primarily logarithmic in nature, were observed between signal characteristics and the dimensions of EDM notches which were used to model actual cracks. Sensitivity level and eddy current system choice were observed to significantly affect the eddy current/flaw dimension relationships.

INTRODUCTION

The design of recent jet engines entering USAF service has emphasized increased performance and higher thrust/weight ratios, resulting in higher stresses on rotating components. This in turn has led to the introduction of larger numbers of finite life components. For example, most rotating disks tend to have relatively short low cycle fatigue (LCF) lives. Since these disks have rapidly increased in cost due to design complexities and the use of advanced materials and processing techniques, the cost of maintaining these advanced engines has also escalated dramatically in the same time period. It

240

is imperative then that ways be sought to optimize the useful
service lives of these components.

In this paper an alternate approach, referred to as
Retirement-for-Cause (RFC), is described. It is believed that
this technique, which is based on a fracture mechanics analysis
of the crack propagation phase, can in fact optimize the service life and thereby minimize maintenance costs.

It will also be shown that the successful application of
Retirement-for-Cause is highly dependent on nondestructive
inspection (NDI) capabilities. Furthermore, this NDI requirement is more quantitative than has generally been considered
in the past since it becomes necessary not only to detect the
defect (crack) but also to describe it in considerable detail
and with a high degree of reliability. While the quantitative
aspects of NDI have only recently begun to be explored in detail, data will be presented that indicates that the necessary
degree of quantification appears to be possible (1,2).

RETIREMENT FOR CAUSE AND NDI REQUIREMENTS

Traditionally, components whose life limit is controlled
by fatigue have been designed to a crack initiation criterion.
The component is considered to have failed as soon as a crack
of some finite size, e.g., .031" (.8mm) has formed and the
part is removed from service (3). No attempt is made to utilize the remaining life associated with the crack propagation
phase.

From a safety standpoint, this apporach has been generally very successful since it contains a built-in safety factor
associated with crack propagation. However, for real materials
and for real design situations, lifetimes based on time to
crack initiation tend to be extremely conservative. This may
be seen by reference to Fig. 1, which illustrates the crack
initiation behavior of Inconel 718, a typical nickel-based
superalloy, at 1000°F. Because of the statistical nature of
the crack initiation process, there is a significant scatter
associated with the number of cycles to initiate a crack at
some given stress level. For design purposes this problem of
materials scatter is usually eliminated by degrading the S/N
curve to a level where the probability of failure, i.e., crack
initiation, becomes low enough to insure structural integrity
of the component. For critical components such as engine
disks, this probability is usually set at 0.1%. Figure 1 shows
a design allowable curve based on this probability of initiation. In service a fatigue limited component would be used

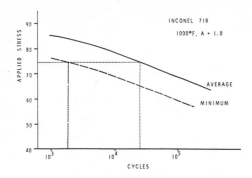

Fig. 1. S/N Diagram for Turbine Disk Alloy, Applied Stress
         vs. Cycles to Crack Initiation.

for the number of cycles permitted by this design curve and
then all such components would be retired. Theoretically, at
this design life only one component of a population of 1000
would have actually initiated a crack and the remaining 999
components would have some undefined useful life remaining.
Reference to Fig. 1 shows that in the case illustrated the
difference between the number of cycles to reach the design
curve and the mean are significantly different and that at
the design limit an average component would have consumed 10%
or less of its potential useful life. However, under an initiation criterion there is no way to utilize this potential
life without accepting a higher probability of failure.

Nevertheless, this additional useful life can be utilized
by adopting a rejection criterion that is based on crack propagation rather than initiation (3). The development of fracture mechanics concepts over the last several years has permitted the degree of predictability for crack propagation
necessary to implement such an approach.

Figure 2 shows the basic retirement for cause concept.
For a given component, the number of cycles, N_c, required to
propagate a crack from an initial size A_o to critical size A_c
can be calculated. This number of cycles, N_c, then becomes
the upper bound for a cracked component to remain in service.
An inspection interval is then established at some fraction
of N_c designated N_I. It can be seen that over this interval
of time no component containing a crack equal to or smaller
than A_o could fail catastrophically.

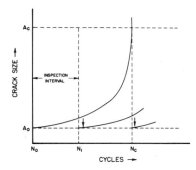

Fig. 2. Crack Growth/NDI Relationship for Retirement for Cause

In operation, components would be inspected at the end of the design life, and only those components containing cracks equal to or greater than A_0 would be retired. All others would be returned to service. After an additional N_I cycles those components would again be inspected and again all cracks larger than A_0 rejected and the remainder returned to service. In this way the crack propagation residual life is continually rezeroed to A_0. By following this approach, components are only rejected for cause (cracks) and the remainder are allowed to operate for the maximum useable time.

It is clear that not all fatigue-limited components may be handled in this way, and that each component must be evaluated individually to determine the economic feasibility. The inspection interval N_I (Fig. 2) must be such that it does not place undue constraints on the operation of the component or that the cost of the necessary tear down and inspection does not negate the advantages of the life extension gained. One thousand cycles of crack propagation may represent many years of service for one component and a fraction of a second for another. It seems unlikely that retirement for cause can be applied to components limited by high cycle fatigue considerations, but for many high cost components limited by low cycle fatigue, such as engine disks, this approach does appear to offer significant economic advantages.

It is also clear that in applying retirement for cause, nondestructive inspection becomes a critical factor. The value of A_0 in Fig. 2 determines the residual life of the component and is limited by the resolution and reliability of the inspection system employed. In many cases the decision as to whether or not retirement for cause can be applied to a component will be predicated upon the ability of available

NDI approaches to detect A_o with sufficient sensitivity and reliability. Because the RFC analysis includes an indepth stress analysis, a component's defect critical locations can be accurately predicted. For this reason, NDI techniques can be selected and refined for a particular area rather than attempting to develop a technique for characterizing the quality of an entire component. This inherently increases the sensitivity of the system to a level where RFC can be utilized. Preliminary crack growth analyses indicate that the detection and elimination of cracks larger than .030-.050" surface length (2c) would provide adequate residual life for the application of RFC to many older disk designs, and this was the crack size of primary interest in the present study. However, it is also recognized that in some of the more advanced designs, using higher strength, lower toughness material, the acceptable level for A_o may be much smaller.

A program of which this research is a part, is looking at the possibility of applying a RFC rejection criterion to a particular turbine disk that is life limited in the bolt hole region due to LCF cracking. A semiautomatic eddy current technique which has undergone extensive development for the inspection of fastener holes in airframes is being reviewed for this engine component application (4). Since cracking around both cooling and bolt holes is becoming an increasing problem, the development of such advanced eddy current techniques is desirable. A particular area of eddy current technology that deserves considerable investigation is that of signal interpretation. In particular, it seems reasonable to suggest that there are various relationships that can be established between signal characteristics and actual flaw dimensions (5). If such correlations could be determined, the information would significantly impact the implementation of reject criteria such as retirement for cause. Inspection techniques could be altered from the basic "go no-go" type to systems which provide opportunity to more easily establish rational reject limits. Another application of such a system is that it would provide an opportunity to accurately determine the shape of a flaw during growth. Both crack growth lives and critical flaw sizes are very sensitive to the aspect ratio of crack, that is, the ratio of the surface length of the crack to the maximum depth. In a RFC analysis for a particular turbine disk, the crack growth lives varied by a factor of 2.5 for the range of flaw aspect ratio, 0.25-0.5. The following discussions will describe research aimed at examining the feasibility of establishing quantitative relationships between eddy current inspection signals and actual crack dimensions.

SIGNAL QUANTIFICATION

Approach

To develop the correlation between signal characteristics and flaw size, the approach originally decided upon was to nondestructively inspect retired, high time disks using the eddy current technique and subsequently to destructively determine the size of detected flaws. Specifically, the cracked areas were to be cut from the disk, oxidized at 1400°F to decorate any fatigue cracks open to the surface, and finally broken open at cryogenic temperatures to reveal the fatigued areas (Fig. 3). The procedure was quite successful for what are considered large cracks (greater than .100" surface length).

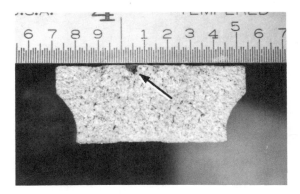

Fig. 3. Service Induced Crack Oxidized to Reveal Fatigued Area.

However, for smaller cracks there were several problems that were encountered. In particular, fatigue cracks in the specific disk material studied. Incoloy 901, generally initiated in several places in the bolt hole radius and the initiation sites were frequently noncoplanar as is illustrated using high resolution dye penetrant (Fig. 4). Thus it was impossible to reveal the entire crack using the above procedure. The only alternative method of documenting the total flaw size was considered to be metallographic sectioning. The amount of effort involved in obtaining sufficient data to establish quantitative relationships, however, seemed prohibitive without further evidence to support the feasibility of the concept.

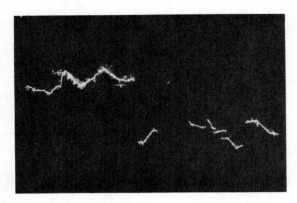

Fig. 4. Penetrant Enhanced Observation of Noncoplanar and
         Multiple Cracking.

As an initial effort to examine the possibilities of
signal quantification and to circumvent the problems with
crack size documentation, it was decided to utilize manufactured notches in the study. Flaws were electron discharge
machined (EDM) in retired disks that were free of any detectable fatigue cracks. Twenty notches were machined with varying aspect ratios in the range of size .011-.282" deep by
.030-.755" in surface length. (Table I) Width of the notches
were from .005 to .007". Although, potentially, the major
value of signal quantification lies in the small crack range,
it was decided to extend the study to the larger flaws in
order to better analyze any physical relationship that might
control the correlation between eddy current signals and flaw
dimensions.

Notches were machined in two basic shapes, rectangular
and semielliptical. Dimensions were determined by making
silicon rubber impressions of each slot and measuring the
lengths on an optical comparator (Fig. 5). Photos were taken
of each impression and the area of each notch was measured.

                Equipment and Experimental Procedures

Two different eddy current systems were used for this
research. System A, shown in Fig. 6, consisted of an Automation Industries Model EM3300 eddy current instrument,
Tektronix oscilloscope and Mosely X-Y recorder. The EM3300
vertical output was AC coupled to the oscilloscope so that
constant or DC signals were not recorded. The oscilloscope

TABLE I

EDM NOTCH DIMENSIONS

Semielliptical		Rectangular	
Length (in)	Depth (in)	Length (in)	Depth (in)
.037	.011	.030	.013
.047	.019	.030	.015
.076	.027	.051	.015
.118	.044	.062	.015
.236	.066	.081	.029
.300	.097	.111	.036
.461	.138	.209	.060
.603	.208	.303	.100
.671	.282	.485	.127
		.755	.182

Fig. 5.   Silicon Rubber Replicas of EDM Notches

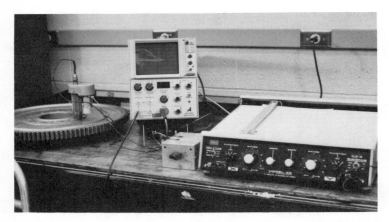

Fig. 6.   Eddy Current System A and Semiautomatic Rotating
Scanner in Place on Turbine Disk.

provided an output proportional to its vertical signal which
was recorded on the X-Y recorder. A typical signal recording
from an EDM notch is shown in Fig. 7. The test coil was
electronically balanced with a reference probe after the test
probe was inserted in an unflawed hole. Probe lift-off correction was made by applying finger pressure to the probe and
adjusting the impedance plane phase control such that any indication due to lift-off was in the horizontal direction only,
thus the only vertical signals recorded were those due to conductivity changes such as those produced by passing the coil
over a discontinuity. Experimentation was conducted at two

sensitivities (gain settings). The excitation frequency was fixed at 500 kHz.

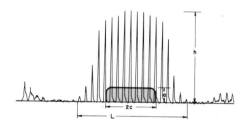

Fig. 7. Typical Eddy Current Signal from EDM Notch (Super-imposed).

The Gulton Industries Model ND2 was the basis of System B. This system used the Magnaflux ED520 eddy current unit which outputs its signal (AC coupled) to a Techni-rite strip chart recorder. The test coil was balanced electronically and lift-off corrections were accomplished by adjusting the coil excitation frequency. The nominal operating frequency was 100-200 kHz. Various sensitivity levels were employed with system B by adjusting the strip chart recorder gain settings. Signals from system A and B were recorded without electronic filtering, other than the high pass AC coupling.

A semiautomatic scanner (Fig. 6) was used to rotate the eddy current probe in a spiral manner through the disk bolt holes. The probe extended axially .025" with each revolution. Scan speeds were held constant at one revolution per second. An absolute eddy current coil embedded in a .625" diameter teflon probe was used for all recordings.

Two scans were made for each bolt hole inspection with an initial coil location for each 180° apart. This effectively increased the resolution of each scan to approximately .0125" per revolution. All recordings were made with the probe extending into the bolt hole. Flaw signals were recorded on hard copy and then were digitized using a Hewlett Parkard 9820, desk top calculator /digitizer. In addition to measuring the maximum signal height, h, and the number of vertical deflections, L, the calculator numerically integrated the digitized signal to obtain the signal area A_{EC}.

Results and Discussion

The experiment was designed to investigate how three basic characteristics of the eddy current signal varied with specific changes in flaw shape using two eddy current systems. The relationships of interest (Fig. 7) were the variation of eddy current signal area, A_{EC}, with notch surface area, A_N; changes in eddy current maximum signal height, h, with notch depth, a; and the number of vertical eddy current signal deflections, L, with notch surface length, 2C.

The experimental results are shown in Figs. 8-12 for eddy current system A. The variation of maximum signal height, h, is plotted versus notch depth, a, in Fig. 8 for both shapes of notches. There appeared to be a logarithmic increase in the maximum height of the signal as the notch depth was increased with minimal effect due to notch geometry. This logarithmic behavior was attributed to the well known skin effect phenomenon which causes the eddy currents to be concentrated near the surface adjacent to the coil and to decrease exponentially with depth (5).

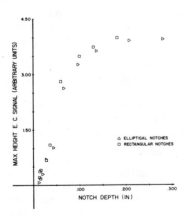

Fig. 8. System A: h vs. a for Two Notch Geometries.

Additional data were plotted in Fig. 9 for the notch depth/maximum peak height relationship. In this case, the

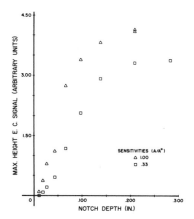

Fig. 9.   System A: h vs. a for Two Sensitivity Levels.

data represented readings for the elliptical notches at two
sensitivities using system A. The sensitivity level was described by the normalizing factor (A/A*) where A is the eddy
current signal area for the particular sensitivity setting and
A* is the signal area for the maximum sensitivity used in these
experiments. As (A/A*) increased, the sensitivity increased.
This figure illustrates that at lower sensitivity levels the
same logarithmic trend was apparent and the saturation level
was not single valued.

A monotonically increasing relationship was also observed
when eddy current vertical deflections were plotted versus
notch surface length (Fig. 10). In this case, however, there
appeared to be a significant geometry effect. It was believed
that the variance in response can be attributed to the differences in aspect ratio between the two sets of notches. It was
also observed that for small notches a large number of vertical
peaks were recorded. This is demonstrated in Fig. 10 by an
apparent nonzero ordinal intercept for the rectangular notches
and was due to the large area or aperture over which the probe
senses. The actual sensing diameter of the probe was estimated
to be 0.22-0.24" (5.5-6.0mm) for this particular sensitivity
level. This large sensing diameter also caused the number of
vertical indications, L, to saturate as the notch surfaces
length neared the edge of the bolt hole. Subtracting the aperture diameter, 0.23" (6mm), from the specimen thickness, 0.83"
(21mm), one would expect the number of vertical indications, L,
to increase linearly with surface length in the crack range
0<2C<0.6" (15.2mm). This appeared to be approximately the case
in this investigation.

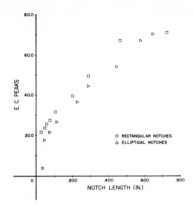

Fig. 10.   System A: L vs. 2c for Two Notch geometries.

Plotting similar data for two different sensitivity levels, Fig. 11 illustrates the slight shift in the relationship as the sensitivity was altered.   Of more interest, however, were the nonlinear points in the lower crack range (.037"-.047" surface length).   Based on the majority of data for this relationship, it is believed that these points deviated from linearity because of the lack of system sensitivity in this crack range. This lack of sensitivity was due to a low signal to noise ratio for these data points.   It can be seen that increasing the sensitivity from (A/A*) of .33 to 1.0 reduced the scatter.

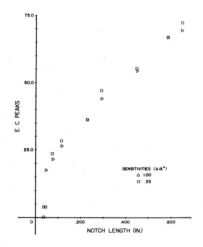

Fig. 11.   System A: L vs. 2c for Two Sensitivity Levels.

The relationship between the area of the eddy current signal, A_{EC}, and the surface area of the notch A_N is shown in Fig. 12. Here, again there was a monotonically increasing

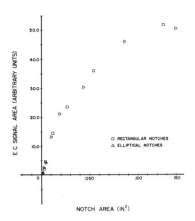

Fig. 12.   System A: A_{EC} vs. A_N for Two Notch Geometries.

function which appeared to be independent of notch geometry. The relationship reached a saturation value which was due to the exponential decrease in current density with depth of penetration.

The experimental results utilizing eddy current System B are presented in Figs. 13-15. These data represent a range of different sensitivity levels, but only for the rectangular notches. In general, the results for System B follow the same trends as those reported for System A. The eddy current signal height, h, versus notch depth, a, relationship for varying sensitivity levels is presented in Fig. 13. Saturation was much more apparent for these results when compared with those results for System A (Fig. 9). This difference was attributed mainly to recorder saturation, and, thus, the logarithmic relationship which would be expected was truncated. One can also see in Fig. 13 that as the sensitivity of the system was increased, upper range saturation became more of a problem. In addition, there appeared to be a lower bound saturation for System B; no explanation of this behavior is apparent.

Eddy current signal area is plotted versus notch area in Fig. 14 for several sensitivity settings. As with System A, the relationships were relatively well behaved monotonically increasing functions. Sensitivity of the eddy current system, again, significantly affected the signal.

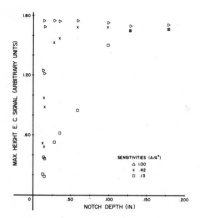

Fig. 13. System B: h vs. a for Three Sensitivity Levels.

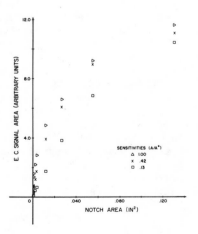

Fig. 14. System B: A_{EC} vs. A_N for Three Sensitivity Levels.

Finally, the number of vertical eddy current indications versus notch surface length relationships for system B are presented in Fig. 15. Regression analyses were performed to determine the degree of linearity for the three sensitivities over various ranges of crack size. The standard error of these best fits relations are plotted in Fig. 16 versus the sensitivity parameter, (A/A*). The results indicated that for the smaller notch length range the standard error decreased slightly as the sensitivity increased. In the larger notch range, however, the relationships deviated significantly from

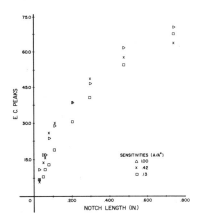

Fig. 15.   System B: L vs. 2c for Three Sensitivity Levels.

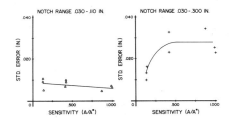

Fig. 16.   System B: Standard Error from Linear Regression
Fit Over Two Notch Size Ranges vs. Sensitivity Level.

linearity as the sensitivity was increased. This can be attributed to the saturation effect observed as probe approaches the edge of the bolt hole.

GENERAL DISCUSSION

The findings from this program demonstrated that relatively well behaved functional relationships exist between notch size and various characteristics of the related eddy current signal. Although the eddy current response from actual fatigue cracks may well differ from the machined notch response, it seems reasonable to assume that relationships similar to those reported herein could be developed between eddy current signals and actual fatigue cracks.

It has been demonstrated that the eddy current inspection technique is sensitive enough to readily detect small flaws. This technique has been shown in this and preliminary work to successfully detect fatigue cracks .030" by .040" and EDM notches as small as .010" by .015". Smaller fatigue cracks were detected; but due to the previously mentioned difficulties the actual size could not be measured. These results support the conclusion that RFC is technically feasible from an NDI standpoint for those older design disks which have demonstrated adequate crack propagation time from this size of detectable flaw.

This research has indicated particular variables that may be important in defining signal/flaw dimension relationships for fatigue cracks. It has been shown that the sensitivity range of the system can significantly affect the type of signal response. In particular, the results suggest that there is an optimum sensitivity for a particular instrument and range of crack sizes. The sensitivity has been defined by the parameter A/A^* and must be large enough to insure detection in the range desired so as to avoid a situation where smaller flaws cannot be discriminated, Fig. 11, without being so large as to cause premature saturation. It should also be noted that the increase in resolution (decreased standard error) for an increase in instrument sensitivity is not large compared to the penalties incurred when saturation is developed (Fig. 16).

Based on this work, it should be expected that each eddy current system will have its own characteristics and, therefore, limitations. Specifically, it is likely that the detection levels and saturation points will differ considerably from system to system. In addition to the problem of system to system variability, is the issue of probe response for a given system. Although they have not been treated herein, the variability of probe response, the effect of probe wear and the interaction of surface condition with signal response are additional inspection parameters which should be documented in order to increase the reliability of the system.

CONCLUSIONS

The observation of consistant relationships between eddy current signal characteristics and notch dimensions indicate that quantitative NDI for fatigue cracks is feasible.

Both the sensitivity level and the choice of a particular eddy current system are important variables to be considered when attempting to establish practical correlations between signal and flaw.

Using the semiautomated eddy current technique, small flaws in the size range of interest in this paper could be readily detected. It is concluded, then, that from an NDI viewpoint RFC is a viable concept.

ACKNOWLEDGEMENT

The authors would like to thank Mr. G. Hardy, AFML/MXA, for this helpful suggestions and consultations.

REFERENCES

(1) D.O. Thompson, "Proceedings of the ARPA/AFML Review of Quantiative NDE", AFML-TR-75-212, January 1976.

(2) P.F. Packman, R.M. Stockton, J.M. Larsen, "Characterization and Measurement of Defects in the Vicinity of Fastener Holes by Nondestructive Inspection", AFOSR-TR-76-0400, Air Force Office of Scientific Research, Bolling AFB, Washington, D.C.

(3) S.A. Sattar and C.V. Sundt, "Gas Turbine Engine Disk Cyclic Life Prediction", Journal of Aircraft, Vol. 12(4), April 1975, p. 360-365.

258 / J. Allison, J. Hyzak, W. Reimann

(4)   A.P. Rogel, "Automatic Eddy Current System for Detection
of Fastener Hole Cracks", Technical Report, Directorate of
Material Management, USAF, McClellan AFB, California, Oct 1971.

(5)   H.L. Libby,"Introduction to Electromagnetic Nondestructive
Test Methods", Wiley-Interscience, New York, 1971.

EVALUATION OF ULTRASONICS AND OPTIMIZED RADIOGRAPHY FOR 2219-T87 ALUMINUM WELDMENTS

W. N. Clotfelter
J. M. Hoop

Marshall Space Flight Center

Quantitative flaw size data are necessary to make reliable decisions on weld acceptability. Nondestructive inspections are usually qualitative in nature and less effective in measuring defect size. The ultrasonic studies described in this report are specifically directed toward the quantitative measurement of randomly located defects previously found in aluminum welds with radiography or with dye penetrants. Experimental radiographic studies were also made to optimize techniques for welds of the thickness range to be used in fabricating the External Tank of the Space Shuttle.

INTRODUCTION

Initially, all welds in the Space Shuttle External Tank will be inspected radiographically and with dye penetrants. As manufacturing experience is obtained and probable flaw distribution patterns emerge, the percentage of welds to be radiographically inspected will be reduced. Quantitative crack size data are necessary to make reliable decisions on weld acceptability. This is especially true when fracture mechanics technology is used as a guide for weld acceptability. Consequently, any defects located by radiography or with penetrants must be reevaluated and carefully measured to obtain size data. Previous work has demonstrated that ultrasonics has high potential for this application. Therefore, a major objective of the work described in

this paper was to demonstrate the utility and reliability
of ultrasonics as a means of assessing the defect content
of aluminum weldments. Since this work is directed
toward the measurement of randomly located defects
found by other inspection methods, portable ultrasonic
instrumentation was selected and manual techniques were
developed. A second objective was to optimize basic
radiography for nondestructively evaluating 2219 aluminum
weldments of a specific thickness range.

The first objective was accomplished by developing
specialized ultrasonic techniques and utilizing them to
nondestructively evaluate weldment flaws. Subsequently,
the specimens were radiographed and destructively
evaluated. A correlation of all test results demonstrated
the current utility and the future potential of ultrasonics
as a weldment evaluation tool. The second objective was
realized by relating the longest apparent crack length
measured in radiographs obtained using a range of film
exposure parameters for each weld specimen. The
particular combination of exposure parameters that
produced the radiograph having the longest apparent
crack length was considered best.

ULTRASONIC FLAW MEASUREMENT TECHNIQUES

Pulse Echo

A Krautkramer Model USK-5 miniature flaw
detector and a MWB 70° shear wave angle beam ultrasonic transducer were used for this application. As
previously stated, this work is directed toward the manual
measurement of randomly located flaws previously found
with radiography or dye penetrants, so portability is
important. The 70° shear wave transducer was selected
since smaller angles result in greater weld bead reflections and larger angles increase difficulties with surface
effects. A commercially available couplant, Exosen 7,
sold by Aerotech, Inc. was used because it proved to be
superior to oils, greases, and other couplants evaluated
with respect to consistency of coupling and energy transfer. Transducer location and orientation with respect to
welds being evaluated are illustrated in Figure 1.

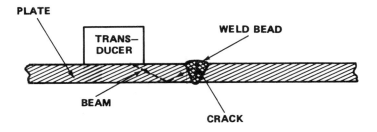

Fig. 1. Shear Wave Crack Measurement Technique.

The ultrasonic instrumentation was calibrated with the aid of a 1 mm "side drilled hole" of an IIT-IIW calibration block. As illustrated in Figure 2, this is simply a round hole drilled perpendicular to the sound beam path. This type of reference is easily manufactured, is reproducible, and reflections from the hole approximate reflections from typical weld defects more closely than those from a flat bottom hole. Figure 2 illustrates the alignment of the MWB 70° transducer used to maximize the echo from the hole. The calibration block must be made of the same material as the item to be tested, and the same couplant as that to be used in the actual test must be used for calibration.

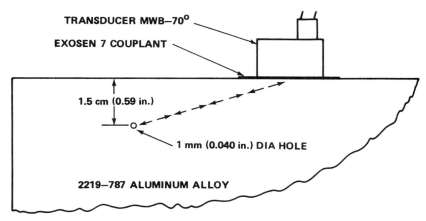

Fig. 2. IIT-IIW Test Block and Method of Calibration.

Test panels were prepared by butt welding 2219-T87 aluminum in 0.32 cm (0.125 in.) and 0.64 cm (0.250 in.) thicknesses using the tungsten inert gas welding process. The panels were intentionally prepared defective by using contaminants. These panels were radiographically and ultrasonically evaluated at 4 MHz with a contact transducer. Subsequently, they were destructively evaluated by sawing them into 4 in. lengths and breaking them with a tensile machine. The actual defects were then measured under a microscope. These data, presented in Tables 1 and 2, were compared to radiographic and ultrasonic data previously obtained.

Of the observed defects, 74 percent were detected radiographically and 81 percent were detected ultrasonically. All defects not detected ultrasonically and 71 percent of those not detected radiographically were an unusual type of porosity. This undetected "porosity" was rather flat and filled with foreign matter that contained very small voids. The difficulty in detecting this condition is well understood. Ultrasonics will not sense an inclusion unless there is a separation between it and the principal material or unless there is a significant difference between the acoustic impedance of the two materials. Furthermore, X-radiation will not sense an inclusion having a density near that of the principal material. However, as previously stated, these weldments were purposely made defective by applying contaminants to the unwelded aluminum plates, so this particular type of porosity is not likely to occur in flight hardware. The results do depict the basic physics and limitations of both inspection methods. Other defects missed by radiography were insufficient fusion, which was oriented at such an angle with respect to the incident X-radiation as to preclude detection.

It should be emphasized that weld beads of all the specimens discussed to this point were unshaved. This unshaved condition constitutes a major limitation to the effectiveness of any nondestructive evaluation method. This is especially true for an ultrasonic method which,

Table 1. Comparison of Nondestructive and Destructive Measurements of Weld Defects for Specimen 1.

No.	Type	Actual Size, cm (in.)	Indicated Size, cm (in.)	
			Radiographic	Ultrasonic
1	Por. & Incl.	0.46 (0.18)	0.51 (0.20)	0.76 (0.30)
2	Porosity	0.18 (0.07)	0.08 (0.03)	0.25 (0.10)
3	IP & Por.	2.31 (0.91)	2.54 (1.00)	2.54 (1.00)
4	Por. & Incl.	0.25 (0.10)	0.21 (0.08)	0.25 (0.10)
5	Porosity	0.13 (0.05)	0.13 (0.05)	0.25 (0.10)
6	Por. & Incl.	0.41 (0.16)	0.30 (0.12)	0.25 (0.10)
7	Porosity	0.36 (0.14)	0.21 (0.08)	0.25 (0.10)
8	IP & Por.	1.93 (0.76)	0.56, 1.27 (0.22, 0.50)	1.27 (0.50)
9	IP & Por.	2.00 (0.79)	0.33, 0.76 (0.13, 0.30)	0.25, 0.76 (0.10, 0.30)
10	IF & Por.	0.84 (0.33)	0.76 (0.30)	0.76 (0.30)
11	IF & Por.	0.56 (0.22)	0.25 (0.10)	0.25 (0.10)
12	IF & Por.	6.10 (2.39)	3.68, 0.25 (1.45, 0.1)	3.80, 0.25 (1.50, 0.1)
13	IF	1.10 (0.42)	ND	Assumed to be bead
14	IF	2.58 (1.02)	ND	indications

Notes: Por. & Incl. - Porosity and Inclusion, IP - Insufficient Penetration, IF - Insufficient Fusion, ND - Not Detected.

Table 2. Comparison of Nondestructive and Destructive
Measurements of Weld Defects for Specimen 2.

No.	Type	Actual Size, cm (in.)	Indicated Size, cm (in.)	
			Radiographic	Ultrasonic
1	Porosity	0.41 (0.16)	ND	ND
2	Porosity	0.23 (0.09)	ND	ND
3	IF & Por.	0.79 (0.31)	0.76 (0.30)	0.51 (0.20)
4	IF & Por.	0.71 (0.28)	0.51 (0.20)	1.02 (0.40)
5	Porosity	0.18 (0.07)	0.30 (0.12)	0.25 (0.10)
6	Porosity	0.15 (0.06)	ND	ND
7	Por. & Incl.	0.48 (0.19)	0.46 (0.18)	2.04 (0.80)
8	IF & Por.	0.79 (0.31)	0.76 (0.30)	1.78 (0.70)
9	IF & Por.	0.71 (0.28)	0.76 (0.30)	0.76 (0.30)
10	IF & Por.	0.56 (0.22)	0.51 (0.20)	0.25 (0.10)
11	IF & Por.	0.53 (0.21)	0.76 (0.30)	0.25 (0.10)
12	Porosity	0.05 (0.02)	ND	ND
13	Porosity	0.23 (0.09)	ND	ND

Notes: Por. & Incl. - Porosity and Inclusion, IP - Insufficient Penetration, IF - Insufficient
Fusion, ND - Not Detected.

with the limitation, consistently detects a higher percentage
of defects in welds than any other method. Shaving weld
beads obviously would improve the detectability of porosity
filled with foreign matter as well as other types of defects.
For example, Figure 3 shows what can be accomplished by
shaving weld beads. A 10 MHz pulse echo, back-reflection
technique was used to obtain a mirror image of imperfec-
tions shown in a radiograph. Obviously the radiograph
was made before the bead was shaved. This improved
capability, along with the inherent capability of ultrasonics
to detect cracks and crack-like defects, provides a very
effective tool for the nondestructive evaluation of weldments.

Pitch and Catch

Cracks transverse to weld beads are not as readily
detected by the angle beam, pulse echo technique, illustrated
in Figure 1, as cracks aligned in the perpendicular direction
or at some intermediate angle to the ultrasonic beam. The
technique depicted in Figure 4 is more sensitive to transverse cracks and results in a higher signal-to-noise ratio
than pulse echo testing. In this pitch and catch technique,
the transmitting transducer "T" directs an ultrasonic beam
at an angle to the crack from which it is reflected to the
receiving transducer "R". The path of the beam through
the plate depends on the plate thickness. If the thickness
is comparable to the beam size, the plate will be essentially
filled with sound and a crack can be detected on either side
of the plate. This was found to be true of 0.64 cm (0.250 in.)
thick aluminum using the miniature Krautkramer transducers.

An attempt to apply this technique by holding a
transducer in each hand was cumbersome and tedious.
A fixture was needed to hold the transducers in a selected
angular alignment and to provide a certain freedom of
motion to allow the transducers to seat themselves on a
test specimen. Such a fixture was designed and fabricated.
Its major features are illustrated in Figure 5. The transducers are held by flanges that are free to rotate about the
points of full dog set screws. These are mounted in

Fig. 3. An Ultrasonic C-Scan of a Weld Compared to a Radiograph of the Weld.

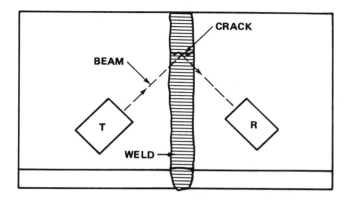

Fig. 4. Transverse Crack Detection Technique.

brackets that are free to rotate about the ends of a threaded bolt that is bent 90°. In addition to accomplishing other objectives, the bend in the bolt serves as an indicator of crack location. Any cracks detected will be just beyond and at the center of the bend.

A limited amount of testing has been accomplished with the transverse crack detection technique by utilizing the special fixture described. Initial results were encouraging since transverse cracks and inclusions were detected. However, additional work is required to determine the utility and reliability of the technique as a tool for detecting and assessing cracks not located by conventional ultrasonic techniques.

MEASUREMENT OF CRACK DEPTH

Ultrasonic techniques for measuring the depth of cracks in weldments of thin, 0.317 to 1.27 cm (0.125 to 0.5 in.), material are in the experimental stage. Prior to the current project, a pitch and catch shadowing technique utilizing the immersion mode of ultrasonic testing was demonstrated. It worked well but is not suitable for the manual evaluation of randomly located cracks. A recently developed manual technique suitable for this application is depicted in Figure 6. Commercially available miniature

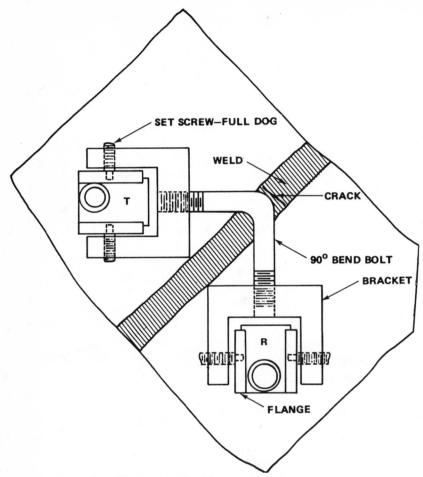

SET SCREW—FULL DOG

WELD

CRACK

90° BEND BOLT

BRACKET

T

R

FLANGE

Fig. 5. Transducer Alignment Fixture for
Use in Detecting Transverse Cracks.

transducers are used with a point contact plastic shoe
bonded to the receiver. A 45° shear wave technique was
selected, since it results in depth and surface shadow
measurements being equal. The transmitting transducer
"T" reflects a beam from the bottom surface in such a
way as to center the tip of the crack in the beam. The
receiving transducer "R" is used to locate the edge of
the shadow. It is not a sharp point but is located by moving

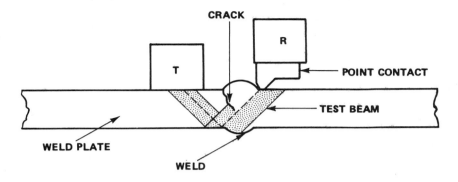

Fig. 6. Manual Ultrasonic Technique for
Measuring Crack Depth.

to essentially zero signal, to a plateau, and then to a
half-amplitude point. A certain amount of skill is required
to locate the end of a crack and the edge of the crack
shadow. Crack depth is approximately equal to the
distance from the tip of the receiving transducer to the
point where the crack reaches the surface. The weld beads
were comparatively smooth, and surprisingly reliable
signals were received when the point contact transducer
was placed on top of them to measure the depth of shallow
cracks. Subsequent to the initial depth measurements,
the beads were milled off and a second set of measurements was made. Then, the material was cut into 0.64 cm
(0.25 in.) wide metallographic specimens, polished, and
etched to reveal crack cross sections. Actual crack
depths were measured with a microscope having a calibrated eyepiece reticule. Results of all measurements, nondestructive and destructive, are presented in Table 3.
The "Position" column identifies incremental measurements of metallographic specimens with respect to the
center of the aluminum plate. Actual crack depth is
compared to the values obtained nondestructively.

In general, the correlation is good for the shaved as
well as for the unshaved weld beads. In both cases an
accuracy of approximately ± 0.10 cm (40 mils) was obtained.
Since the cracks were not vertical, this is considered good.

Table 3. Manual Ultrasonic Crack Depth Measurements.

Specimen: 2219-T87 Weld Panel of 1.27 cm (0.5 in.)
Thickness With Stress Crack in Fusion Line.

Position		Depth, cm (in.)		
		Actual	Measured	
			With Beads	Shaved
Left	12	0	0	0
	11	0	0	0
	10	0	0	0
	9	0	0	0
	8	0.51 (0.20)	0.43 (0.17)	0.23 (0.09)
	7	0.56 (0.22)	0.58 (0.23)	0.48 (0.19)
	6	0.66 (0.26)	0.79 (0.31)	0.51 (0.20)
	5	0.69 (0.27)	0.69 (0.27)	0.58 (0.23)
	4	0.74 (0.29)	0.84 (0.33)	0.69 (0.27)
	3	0.79 (0.31)	0.79 (0.31)	0.63 (0.25)
	2	0.79 (0.31)	0.74 (0.29)	0.81 (0.32)
	1	0.76 (0.30)	0.74 (0.29)	0.71 (0.28)
	0	0.84 (0.33)	0.94 (0.37)	0.79 (0.31)
	1	0.74 (0.29)	0.94 (0.37)	0.69 (0.27)
	2	0.69 (0.27)	0.86 (0.34)	0.71 (0.28)
	3	0.66 (0.26)	0.74 (0.29)	0.56 (0.22)
	4	0.66 (0.26)	0.76 (0.30)	0.53 (0.21)
	5	0.63 (0.25)	0.79 (0.31)	0.71 (0.28)
	6	0.53 (0.21)	0.69 (0.27)	0.53 (0.21)
	7	0.48 (0.19)	0.66 (0.26)	0.43 (0.17)
	8	0.48 (0.19)	0.56 (0.22)	0.28 (0.11)
	9	0.20 (0.08)	0.30 (0.12)	0.10 (0.04)
	10	0.13 (0.05)	0.23 (0.09)	0
	11	0	0.23 (0.09)	0
Right	12	0	0	0

METHOD OF OPTIMIZING RADIOGRAPHIC PARAMETERS FOR WELD EVALUATION

An experienced radiographer knows the approximate values of voltage, amperage, film-source distance, and exposure time for each thickness of aluminum in the 0.317 to 1.27 cm (0.125 to 0.5 in.) thickness range. It is also known that longer exposure times and smaller focal spot sizes will improve film image quality. Usually exposure times for aluminum welds have been held to 1 min or less, and the required 2 percent penetrameter sensitivity is met. However, a reasonable increase in exposure time would be a small price to pay for improved film image quality when critical welds are being evaluated. Thus, effects of exposure time and other radiographic parameters on film image quality have been evaluated so that optimum inspection techniques can be established for each specific material thickness of interest.

If amperage and film source distance are held constant, the effect of increased exposure time is to lower the voltage in order to maintain film density within a readable limit. The effect of amperage is similar to exposure time, namely, to increase the total flux to the film. The film source distance also affects the total flux density, but the practical matter of adequate specimen coverage with the cone of radiation and other considerations limit use of the film source distance as a variable parameter. Therefore, the film source distance and amperage were held essentially constant during these studies. Voltage and exposure time were the major independent variables.

We know that film resolution increases with density, provided the limits of the eye and the intensity level of the viewing screen are not exceeded. Density became the chief dependent variable for initial film evaluation. The relationship of density variations to film exposure parameters is shown in Figure 7. Density is proportional to voltage and exposure time. We also

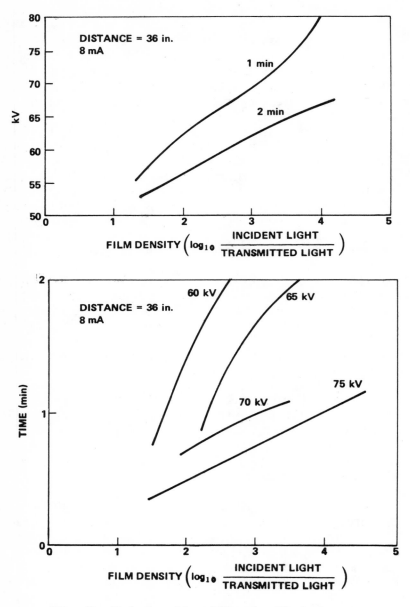

Fig. 7. Relationship of Density Variation to
Film Exposure Parameters .

know that the lowest voltage possible should be used because an increase in voltage increases the percentage of radiation having shorter wavelengths. Short wavelength radiation is more penetrating than long wavelength radiation and results in less contrast in the image of the object being radiographed. Additionally, radiographic film is more sensitive to long wavelengths; thus, the long wavelengths improve image contrast.

Exposure data revealed that the 0.7 mm focal spot was better than any other available size. The X-ray equipment specifications establish the maximum amperage for this focal spot size at 8 to 10 mA. A film source distance of 0.915 m (36 in.) was previously established as the minimum that would give adequate coverage of the test panels. Thus, voltage and exposure time are the only remaining parameters to be varied. Numerous radiographs of weldments obtained by utilizing different combinations of these parameters were made. However, it was necessary to limit the maximum length of exposure time to 2 min. This was based on an estimation of the maximum inspection time that would be acceptable to production personnel.

It was postulated that optimum radiographic parameters are those that produce the greatest defect indications. For example, a crack gets tighter toward the ends and becomes more difficult to detect radiographically or by any other method. Thus, any technique showing maximum length is obviously better than others. Therefore, the apparent length of each defect indication on every film obtained by utilizing various radiographic parameters was carefully measured with the aid of a 7X microscope. The total defect length determined for each exposure condition was then plotted versus the corresponding film density. Curves of this type for a 0.318 cm (0.125 in.) thick aluminum weld are shown in Figure 8. For each exposure time the apparent defect length reaches a maximum as a function of voltage and then recedes as the voltage goes still higher. The combination of parameters yielding the maximum defect length is the optimum technique for this

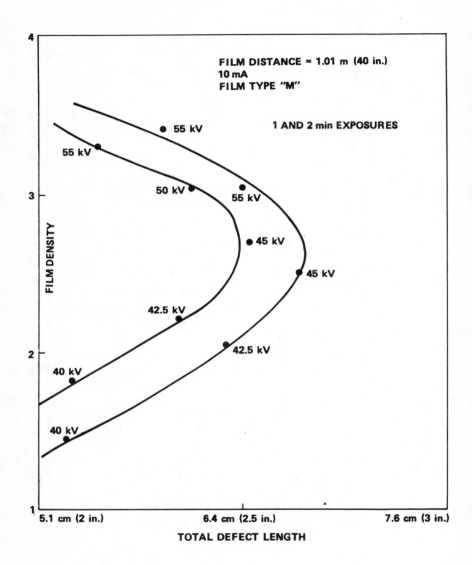

Fig. 8. Selection of Film Exposure Parameters for 0.318 cm (0.125 in.) Thick Aluminum Welds.

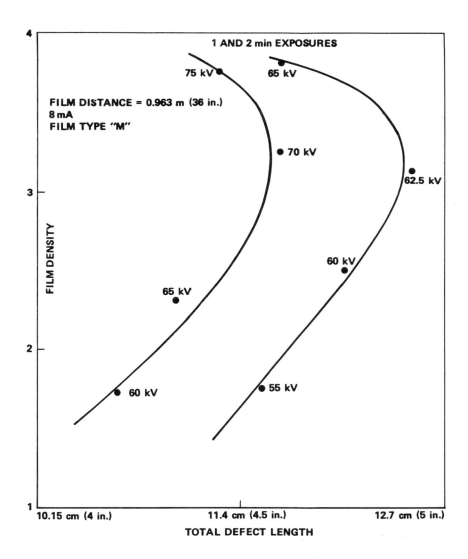

Fig. 9. Selection of Film Exposure Parameters
for 0.99 cm (0.39 in.) Thick Aluminum Welds.

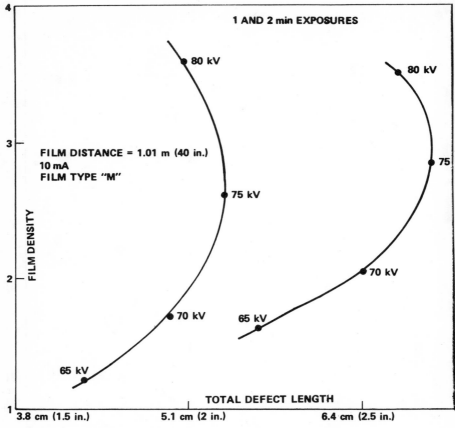

Fig. 10. Selection of Film Exposure Parameters for 1.27 cm (0.50 in.) Thick Aluminum Welds.

as most often applied, are inadequate calibration and interpretation of defect indications. The best way to overcome these problems is to evaluate statistically significant numbers of weld specimens with well calibrated instrumentation and, subsequently, correlate the indications with the real defect as revealed by destructive measurements. This type of analysis is necessary for effective manual ultrasonic testing as well as for highly sophisticated automated ultrasonic testing.

particular material thickness. Similar curves for other thickness values are shown in Figures 9 and 10. It is of interest to note that increases in apparent crack length as exposure time is increased from 1 to 2 min become progressively greater as material thickness increases. It should also be remembered that optimum film exposure parameters will vary to some extent among different X-ray machines.

CONCLUSIONS

Ultrasonic shear waves used in the pulse echo mode provide an acceptable way of measuring the length of weld defects nondestructively, but a pitch and catch shadow technique is required to make accurate crack depth measurements. Weld bead conditions affect the accuracy of both techniques. Ideally, from an inspector's point of view, weld beads should be shaved, but reliable measurements can be made if the bead profile is low and uniform. Flaw size data obtained ultrasonically compare favorably with radiographic data and with real flaw sizes determined by destructive measurements. Thus, the manual pulse echo and shadow techniques described are effective for measuring flaw size and are inexpensive for evaluating a limited number of flaws. This was the major objective of the project.

Results obtained by the inspection of a shaved weldment with a 10 MHz pulse echo, back-reflection technique are a good indication of the potential of ultrasonics for the nondestructive evaluation of welds. A C-scan recording will show inclusions and porosity very much like a radiograph, as well as indications from randomly oriented cracks. This and other ultrasonic techniques can be automated. Therefore, a low cost ultrasonic system can be used instead of radiography to inspect aluminum welds even though the beads are not entirely removed. Although ultrasonic instrumentation can be improved, available systems can be used to obtain weld evaluations equal to or better than those made with radiography. The major shortcomings of ultrasonics,

It has been demonstrated that a careful selection of film exposure parameters for a particular application must be made to obtain optimized flaw detectability. Lower voltage and longer exposure times than those ordinarily used for aluminum welds yield improved results. Radiography and state-of-the-art ultrasonics complement each other, and the combination provides good flaw detection capability.

RECOMMENDATIONS

Four major areas of ultrasonic technology must be improved if its full potential as a quantitative tool for measuring the size of defects in metals is to be realized:

1. Electronic subsystem.

2. Transducer or acoustical subsystem.

3. Calibration procedures for the entire ultrasonic system.

4. Correlation of ultrasonic indications with real defect sizes in materials as obtained by destructive analysis.

Characteristics of available electronic subsystems vary widely. This makes it almost impossible to obtain uniform inspection results from two or more instruments when they are used to evaluate any specimen containing several defects. Furthermore, little would be accomplished by measuring circuit parameters unless provisions are made for adjusting them to meet specified requirements. The variation of transducer characteristics is also significant. Improved quality control of manufacturing procedures, a rigorous initial evaluation of transducer characteristics, and simple, easily applied procedures for periodically checking transducers are necessary to overcome this problem.

Subsequent to the availability of suitable electronics and well characterized transducers, optimized calibration procedures can be developed for each type of inspection problem. Then, one of the most neglected areas of ultrasonic testing can be addressed in a meaningful manner, that is, as illustrated in this report, a systematic correlation of defect indications with real defect sizes as determined by destructive analysis. Statistically significant numbers of specimens containing realistic, naturally occurring defects should be evaluated.

The briefly outlined four point program is highly recommended as the most logical and least costly way of realizing the full potential of ultrasonics as a tool for nondestructive testing. These major areas of ultrasonic technology are being addressed in developmental studies currently being conducted.

A COMPARISON OF NEUTRON AND X-RADIOGRAPHIC
TECHNIQUES AS APPLIED TO AEROSPACE COMPONENTS

Patricia A. Campbell
Bendix Launch Support Division
Kennedy Space Center, Florida

Neutron radiography is a relatively new technique. The majority of industrial applications have been developed within the last ten years. Its primary advantage is that it is a complement to X-radiography. This is because the absorption characteristics of most elements are essentially reversed for neutrons and X-rays; thermal neutrons are attenuated by light elements such as hydrogen or boron and X-rays are attenuated by heavy elements such as lead. This difference allows nondestructive testing of items previously unfeasible or impossible with X-ray, e.g., rocket propellants, aircraft components, radioactive materials, composites, light elements and biological specimens, to be accomplished using neutron radiography.

Most of the original work has been accomplished with nuclear reactors; however, other sources of neutron beams are available, including Van de Graff generators and the recently available Cf-252 isotope. Two disadvantages of neutron source systems are the fixed locations and exposure times for isotope sources measured in hours rather than minutes.

Cf-252 SYSTEM

The system procured by NASA in December 1974 was the IRT Corporation Model CFNR-10, which employs a 10mg source of Cf-252. The construction of the neutron camera is shown in Figure 1. The source is on loan from the Nuclear Regulatory Commission thru the Californium Demonstration Center managed by the

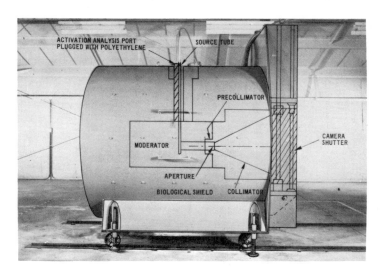

Fig. 1. IRT Corporation Model CFNR-10 camera
showing internal construction.

IRT Corporation. Cf-252 has a half-life of 2.65
years with a fast neutron yield of about 2.3×10^{12}
n/sec/g. The fast neutrons are first moderated to
thermal velocities, then collimated, and finally
detected by an image recorder sensitive to thermal
neutrons. The Cf-252 source is moderated by hydro-
genous materials such as water, paraffin or in this
case high density polyethylene. A precollimator is
fabricated from a material with a high cross section
for thermal neutrons, such as lithium carbonate
enriched with Li^6 isotope, and is located at the
point of peak flux to extract and define the neutron
beam. Filters such as lead or bismuth can be placed
in the precollimator to reduce the gamma radiation
component in the beam. An annular cone loaded with
lithium carbonate powder forms a divergent collimator
which further defines the neutron beam. Borated
water-extended polyester (WEP) and lead are used
for the 16,000 pounds of biological shielding required.

The image recorder in this system utilizes the
direct exposure technique and consists of a converter screen of vapor deposited gadolinium in direct

contact with the film emulsion in a vacuum cassette.
Standard X-radiographic films such as EK-AA, M and
SR are used. The gadolinium absorbs thermal neutrons
and in turn emits soft electrons (70 keV) to expose
the film. The conversion is about 97% efficient.
Also, because the beam contains gamma rays as well
as neutrons, the exposure is to some degree a
composite.

Two image quality indicators, designated E-
545-75, have been accepted by American Society for
Testing Materials for use with neutron radiography.
One is a sensitivity gage (IQI) which provides an
indication of the combined effects of contrast and
unsharpness. This Type A sensitivity indicator is
a double stepwedge of acrylic resin. A series of
.005, .010, .020, and .040 inch steps with holes of
the same depth is attached at 90° to a stepwedge
with 0.031, .062, .125 and .250 inch steps. The
sensitivity level is determined by the number of
consecutive observable holes in the corresponding
thickness of the indicator material as imaged on
film.

The second indicator is a beam purity indicator
(BPI). It is made up of an 8.0mm boron nitride slab
with three holes: one with a thin boron nitride
disc to remove thermal neutrons, one with discs of
lead and boron nitride to remove thermal neutrons
and low energy gamma rays, and one empty hole which
removes scattered neutrons. The boron nitride slab
removes all neutrons and allows the gamma rays to
expose the film behind it. From the film density
at these locations, as well as background density,
measurement of those parameters and radiation components which contribute to film exposure, i.e.,
thermal, epithermal and scattered neutrons and gamma
rays, may be made. The BPI is also used to verify
the reproducibility of the neutron radiographic
system on a day-to-day basis.

APPLICATIONS

The program outline for the two year feasibility
study, KSC Project No. 953-00-00-93, at the Kennedy
Space Center includes establishment of system characteristics, exposure parameters for common metals

and nonmetals, and the study of pyrotechnics,
components, and composite materials such as thermal
protection shield (TPS) panels for the Space Shuttle
Program.

The following are applications of special interest which have been investigated. A comparison
of the radiographs will show the complementary
nature of the two techniques.

Corrosion

A section of 6061-T6 aluminum pipe, 2 inches
in diameter, had been exposed to salt air environmental conditions for about four years at the
Kennedy Space Center. The corrosion products are
readily visible. The X-ray shows no indication of
corrosion, whereas the n-ray has a mottled appearance showing the extent of the corrosion as seen in
Figure 2. The light areas indicate the corrosion
products, mostly $Al(OH)_3$ because of the greater
attenuation of neutrons by hydrogen contained in
$Al(OH)_3$ and water of hydration. An investigation
has been made by IRT Corporation to equate film
density difference to corrosion thickness.

Fig. 2. X- (top) and n-radiographs of corroded
aluminum pipe.

Solenoid Valves

Figure 3 shows radiographs of solenoid valves which were intentionally defected. There is no evidence of soft goods in the X-radiograph. Missing and improper O-rings were noted correctly on the n-ray as discrepancies by comparison to a specification drawing and/or a neutron radiograph of a correctly assembled valve. Insulation stripped from the wiring was not discernable due to small wire gage and its orientation to the film. However, other areas were noted which had not been intentionally defected. A small nylon thread protector was jammed into an allen screw hole located at the bottom of the poppet (arrow 1). Its presence, however, presents no functional complications. The O-rings in the poppet are visible on the n-ray while the teflon back-up rings (arrow 2) are not and neither are visible on X-ray. By noting the position of the O-rings within the groove, it was found that in some locations the rings had been unintentionally reversed - which can present functional complications.

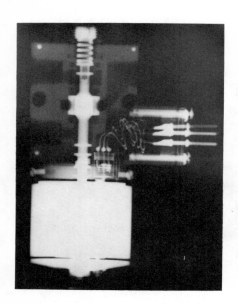 

Fig. 3. X- (left) and n-radiographs of solenoid valves.

Capacitors

A problem arose concerning metallized paper
mylar capacitors which had failed in vibration
testing. The failures were caused by voids in the
supporting wax which fills the capacitors. The
capacitors were radiographed in both side and end
views as seen in Figures 4 and 5 respectively. In
the X-radiographs the dual aluminum housing with
holes for injection of an epoxy bonding material and
the terminals are visible. In the n-radiographs
voids in the wax appear as dark areas. One capacitor
appears off-center in the X-radiograph, Figure 5,
left top, and shows no center core in the n-radiograph (arrow). On disassembly it was found to have
been crushed.

Fig. 4. X- (left) and n-radiographs of side
view of capacitors.

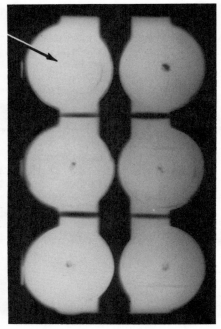

Fig. 5. X- (left) and n-radiographs of end view of capacitors.

Ordnance

Several types of pyrotechnic devices were radiographed with three being of particular note. These are spin rockets and ignitors, shape charges and a stainless steel expandable tube.

Spin Rocket and Ignitors

The spin rocket and ignitors are an excellent example of the complementary nature of the two techniques as can be seen in Figure 6. One main area of interest when X-raying the spin rocket is the star pattern of the propellant in the body. The propellant is inspected for voids or cracks which may cause uneven burning. The star pattern is not visible on an n-ray because the propellant has a high attenuation for neutrons and is not penetrated. The silicone rubber (Silastic 651) nozzle plug and a graphite insert are visible on the n-ray but not on the X-ray. The depth and angle of insertion can be clearly noted. Also, the inhibitor of nylon and epoxy adhesive between the rocket case and the propellant can be seen on the X-ray but not on the n-ray.

In the ignitors, the $B-KNO_3$ propellant pellets are visible on the neutron radiograph (arrow) while pins, bridge wires, etc., must be located by X-radiography.

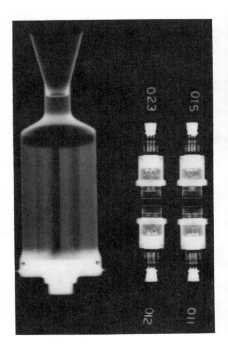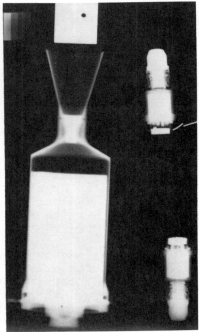

Fig. 6. X- (left) and n-radiographs of spin rocket and ignitors.

288 / P. A. Campbell

Shape Charge

Shape charges, or tension tie cutters, as shown in Figure 7, contain a high explosive material which on firing travels at 20,000 ft/sec. It is basically a mild detonating fuse drawn into a 'V' shape and is used for metal cutting applications in spacecraft separation systems.

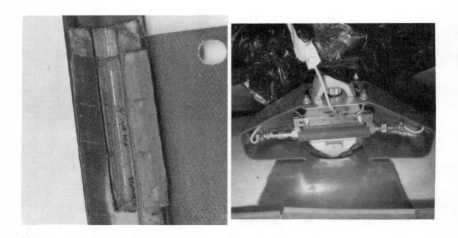

Fig. 7. Left: Shape charge, or tension tie cutter, held in bracket with RTV adhesive. Right: Shape charge assembly which separates Apollo Command Module from Service Module just prior to re-entry.

The two lengths shown in Figure 8 are about 4-1/2" long with a lead sheath around a hexanitro-stilbene (HNS) core. Both ends contain a lead azide booster charge inside an aluminum cap. They are bonded to the aluminum bracket with RTV adhesive. A comparison of the two radiographs shows several differences. On the n-ray, the HNS charge is visible behind the RTV while the end cap is not visible, indicative of the low neutron attenuation of aluminum and lead azide. Visible on the X-ray is the booster charge, the lead sheath covering the HNS, and on close examination, the outline of the aluminum cap.

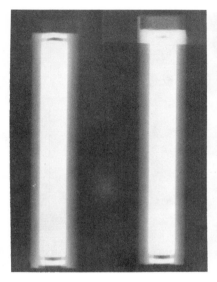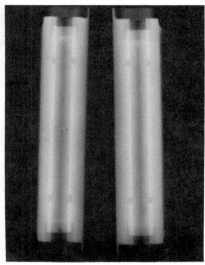

Fig. 8. X- (left) and n-radiographs of two
lengths of shape charge.

Expandable Tube

Expandable tubes originally have an oblong
cross section containing a high explosive core of
10-20 gr/ft of RDX. Initiated by a detonator, the
RDX produces gas and shock to force the tube into
a circular shape thereby breaking a tension link as
on the Skylab meteoroid shield shown in Figure 9.
They are also used to separate solar array systems
for deployment. They are a contained explosive
device which produces no outgassing or shrapnel.

A comparison of the neutron and X-radiographs
of this stainless steel housed expandable tube is
shown in Figure 10. The X-ray shows the 304SS
tubing, the lead sheathing of the center length of
RDX, and an apparent void area around the threads
of the housing of the booster detonator which is
only slightly visible. The n-ray is much more
definitive. It clearly shows the mild detonating
fuse, sections of nonmetallic spacers and the
booster detonator. The void area at the threads is
not empty but is filled with a sealing compound.

Fig. 9. Skylab meteoroid shield with expandable tube pyrotechnic device having broken tension link.

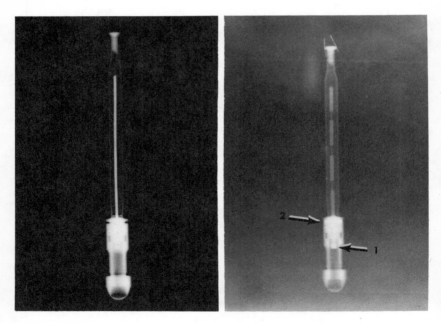

Fig. 10. X- (left) and n-radiographs of expandable tube. N-radiograph shows nonmetallic spacers, booster detonator (arrow 1) and sealant in threads (arrow 2).

CONCLUSIONS

The examples presented here demonstrate that neutron radiography is a valuable complement to X-radiography and supplement to existing NDT radiographic techniques. It is especially so for corrosion detection on aluminum aircraft assemblies, quality control of pyrotechnic devices and missing or misplaced soft goods in components.

A PROGRAM TO STUDY METHODS OF
PLASTIC FRACTURE

J. P. D. Wilkinson
General Electric Company

G. T. Hahn
Battelle Columbus Laboratories

R. E. Smith
Electric Power Research Institute

INTRODUCTION

This paper describes plans for a study to develop a methodology for plastic fracture. Such a fracture mechanics methodology, having application in the plastic regime, is required to assess the margin of safety inherent in nuclear reactor pressure vessels. While linear elastic fracture mechanics and elastic-plastic procedures (such as the J-integral or crack opening displacement) appear to be adequate predictive techniques for some plant conditions, direct application to material under fully plastic deformation is questionable. The steels used to fabricate nuclear pressure vessels will deform significantly at the operating conditions prior to the inception of fracture. Indeed, it is likely that once initiated, flaws will propagate in a stable manner prior to final fracture. In addition, the energy required to effect plastic fracture is much greater than that associated with more brittle failure modes characteristic of low temperature fracture. Both of these characteristics must be embraced in a quantitative plastic fracture methodology. Such a methodology will permit a proper assessment of design safety margins without the need to assume unrealistically conservative boundary conditions which may result from the application of linear elastic fracture mechanics.

The planned program consists of four phases:
First, the selection, through analysis and laboratory experiments, of viable criteria for crack initiation and growth
during plastic fracture. Second, the selection of final criteria for crack behavior through further exhaustive laboratory tests, and the development of a methodology for predicting plastic fracture behavior in structures. Third, the
verification of this methodology through engineering scale
tests. Fourth, the application of the methodology in predicting the behavior of flawed pressure vessels.

The planned program will continue over a period
of 38 months. During the first 18 months, investigators
from General Electric Company and Battelle Columbus
Laboratories will develop the fundamentals of crack initiation and growth under large scale plastic yielding conditions,
leading to criteria for such crack behavior. For the remaining 20 months, the General Electric team will further verify the criteria through both laboratory and large scale tests
and will apply the methodology for plastic fracture so developed. Consultations with J. R. Rice of Brown University and with J. W. Hutchinson of Harvard University will
provide useful input throughout the program.

PROGRAM APPROACH

The initiation and growth of flaws in pressure
vessels under overload conditions is distinguished by a
number of unique features, such as large scale yielding,
three-dimensional structural and flaw configurations, and
failure instabilities that may be controlled by either toughness or plastic flow. In order to develop a broadly applicable methodology of plastic fracture, these features require the following analytical and experimental studies:
development of criteria for crack initiation and growth
under large scale yielding; the use of the finite element
method to describe elastic-plastic behavior of both the
structure and the crack tip region; and extensive experimental studies on laboratory scale and large scale specimens, which attempt to reproduce the pertinent plastic flow
and crack growth phenomena.

294 / J. Wilkinson, G. Hahn, R. Smith

During the initial phases of the program both investigating groups will examine a variety of candidate criteria for crack initiation and growth. In this study, crack initiation is taken to mean the start of crack growth from pre-existing flaws or cracks. For the case of crack initiation, these criteria include the J-integral (1, 2), crack opening displacement (1, 3), and strain amplitude (4), whose presently perceived advantages and disadvantages are shown in Table 1. In the case of crack growth, the criteria to be examined include the J-integral (2), crack tip opening displacement (5), and crack opening angle (5), and some approaches which include the strain amplitude at the crack tip, work done in a crack tip process zone, and a generalized energy release-rate approach. These criteria are listed in Table 2, where the presently perceived advantages and disadvantages of each are shown. It is intended that analysis and testing will complement one another: each specimen configuration will be analyzed through the finite element method in order to predict its experimental behavior. As analysis and testing proceed, the range of candidate criteria will be narrowed to one or two, and finally, during the sixteenth month of the program, a decision will be made as to which criteria will be adopted for describing the process of crack initiation and growth during the remainder of the program.

The basic material to be used in the General Electric program is a single heat of reactor grade A533 Grade B Class 1 steel, purchased from Lukens Steel Company in the form of a plate of size 4.5 m (178 in.) square and 0.2 m (8 in.) thick. The Battelle study will make use of this heat, but the bulk of the work will be carried out on two alloys with yield strength-to-toughness ratios about five times larger than A533B. These "toughness scaled" alloys, in the form of small 6.4 mm to 25 mm (0.25 in. to 1.0 in.) thick laboratory specimens, are calculated to behave like the full scale vessel in the sense that they will display the same failure modes and instabilities as the 0.2 m (8 in.) thick A533B steel wall. In addition, for the General Electric study, A508 Class 2 steel forgings will be used to develop crack initiation and growth data, and tests will also be carried out on specimens containing weld metal cut from commercial weldments.

Table 1: Potential Crack Initiation Criteria

Suitable Near-Tip Parameters	Advantages	Disadvantages
J-Integral (1, 2)	• characterizes near-tip field • easy to calculate and measure	• valid for deformation theory • essentially proportional loading • may be invalid for unloading
Crack Opening Displacement (3) Crack Tip Opening Displacement (1, 3)	• not tied to a particular plasticity theory • relatable to micromechanisms of fracture • characterizes near-tip field	• has precise meaning only for perfect plasticity • computation requires accurate crack tip modeling • difficult to measure
Strain Amplitude (4)	• not tied to a particular plasticity theory • characterizes near-tip field	• difficult to calculate and measure
Work Done in a Crack Tip Process Zone	• relatable to LEFM parameters and micromechanisms of fracture • easy to calculate	• rests on a determination of relevant size of process zone

Table 2: Potential Crack Growth Criteria

Criterion	Advantages	Disadvantages
J-Integral (2)	• easy to calculate and measure	• unloading makes validity doubtful
Crack Tip Opening Displacement (5)	• relatable to micromechanisms of fracture	• difficult to measure
Crack Opening Angle (5)	• easy to measure	• physical basis not established
Strain Amplitude, H-Integral	• characterizes near-tip field for growing crack	• difficult to calculate and measure
Work Done in a Crack Tip Process Zone	• relatable to LEFM parameters and micromechanisms of fracture • easy to calculate	• rests on a determination of relevant size of process zone
Energy Balance	• extension of G-criterion to ductile crack growth	• difficult to calculate

Whereas the global view of the program approach, particularly in its initial phases, is as described above, the details of the approaches to be pursued by General Electric and Battelle investigators lay differing emphasis on one or another criteria, and will also pursue differing experimental philosophies. Therefore, in describing the program details in what follows, the Battelle program will be described first, followed by the General Electric program. It should be emphasized, however, that continuous technical information exchanges are planned, and the developed methodology will contain features from both programs. In addition, specific attention has been paid during the program planning to make these programs complementary and to avoid duplication of effort.

THE BATTELLE PROGRAM

The Battelle program assigns the highest priority to the solution of problems involving deep part-through (or buried) cracks in the vessel wall. This type of defect is more likely to escape detection than a through-the-wall crack and poses the important leak versus break question. At the same time, the part-through geometry presents rather special and difficult analysis and verification problems, and is not the logical starting point for developing a plastic fracture methodology. For this reason, the preliminary plans envisioned for the program would begin by focusing on a crack extension in through-cracked center cracked panels -- a simpler configuration. Attention would next be shifted to the more difficult part-through cracked plate, and finally to double cantilever beam specimens (similar in shape to the compact tension specimen) which are more economical for routine material property measurements.

Finite Element Analyses and Crack Criteria

Finite element methods can handle the currently used crack extension criteria in terms of G_c or $R(a)$, J_c and COD_c, but each of these parameters presents some difficulties when they are applied in the presence of large scale plasticity. The quantities G, $R(a)$, and J_c, which reflect the rate of plastic energy expenditure with crack

298 / J. Wilkinson, G. Hahn, R. Smith

extension, will be influenced by boundaries of structure
when the boundaries interact with crack tip plastic zones.
Consequently, G_c, R(a), and J_c will not be rigorously geometry independent material properties in the case of large
scale yielding. The quantity J_c loses its meaning when
plastic (as opposed to nonlinear elastic) material is unload-
ed during stable crack growth. Finally, the quantity COD_c,
a valid measure of cracking resistance after general yielding, is difficult to define and measure when the crack begins
to grow. For this reason, special emphasis will be given to
developing alternative criteria. One promising candidate is
a generalized energy release rate approach involving R, the
energy dissipated in an "excluded" region within the plastic
zone. The R_c will be calculated and evaluated together with
G_c, R(a), J_c(a), COD_c(a), COA_c(a) and other candidate
parameters emerging from the General Electric program
with the aim of identifying the criterion or parameter best
suited for the pressure vessel problem. An elastic-plastic
large deformation, finite element model is to be constructed
which is capable of analyzing stable crack growth and plastic flow and fracture instabilities in essentially two-dimensional geometries. The models will then be used to analyze the results of the experiments, particularly the validity
of fracture criteria based on R_c, COD_c, R(a), J_c(a),
COA_c(a), and other concepts emerging from the General
Electric program.

The stress analysis computer program will be
used to conduct a study of fracture in a center cracked panel
(CCP). Experimental data will be used to describe plate
geometries and material constitutive behavior. First, to
verify the analysis, comparisons between predicted and
measured properties of the CCP for tensile loading will be
made. Comparisons will include load-displacement curves
and strains at selected points. After establishing that the
analysis is capable of predicting basic mechanics of behavior of the CCP configuration for the materials used in the
experiments, comparisons of crack propagation measurements and calculations will be made. These will involve
utilization of the proposed fracture criteria and the estimation of a critical value of a material property that governs
crack propagation. Of particular interest will be the model-

ing of local plastic instability and comparisons of crack propagation versus instability for this configuration. Similar studies will be made of the part-through cracked plate configuration and of the double cantilever beam specimens. In addition, simplified analyses that approximate the key features of the finite element models will be identified. Further verification of the analytical procedures and criteria is desirable, particularly for A533B in section sizes comparable to reactor vessel wall thickness. This will be done by constructing finite element models of the Southwest Research Institute (SwRI) small and intermediate A533B flat plate experiments (6), and possibly other well documented experiments in the literature. The aim of the calculations will be to reproduce the load-extension curves, strain distribution and fracture stresses reported for the test pieces using independently derived flow and fracture data, and the fracture criterion determined during this study. Since the SwRI experiments involve three-dimensional part-through crack configurations, efforts will be made to structure the two-dimensional model or approximate the actual three-dimensional cracked test pieces. The necessary experimental data will be extracted from the literature.

Verification of Analytical Procedures and Criteria

The acquisition of measurements of deformation, crack extension and instability for direct comparison with the finite element models and the verification of the different crack extension criteria is essential. Two approaches will be used. The first approach involves systematic measurements of part-through and through cracked plates of "toughness-scaled" materials, and the A533B steel. The second approach provides for data reduction of the SwRI, A533B flat plate experiments and other measurements in the literature for direct comparisons with finite element analyses.

Systematic measurements of the deformation, stable crack growth and instability of 0.25 in. to 1 in. thick, part-through and through cracked flat plates and panels will be carried out on "toughness-scaled" materials and A533B steel. The "toughness-scaled" materials will be selected from among aluminum alloys, high strength steel and A533B, heat treated or tested* to produce a toughness-to-yield ratio, (K_{Ic}/σ_Y), consistent with both the scale of the laboratory experiment, and the relationship between flaw size, plastic zone size, and wall thickness of the full scale vessel. This will be accomplished by reproducing in the laboratory experiment, the relative toughness ratio, $(K_{Ic}/\sigma_Y)^2 \, 1/t$, of the shelf-level reactor vessel wall, typically** $(K_{Ic}^Y/\sigma_Y)^2 \, 1/t \approx 2$. Experiments on 0.25 in. thick material would therefore be performed on a "toughness-scaled" material with a toughness-to-yield ratio $K_{Ic}/\sigma_Y = 0.7\sqrt{in}$., compared to the value $\sim 4\sqrt{in}$. displayed by shelf-level A533B steel. This strategy will make it possible to reduce the scale of laboratory experiments (relative to the reactor vessel) by a factor of about 10 to 50, and still observe the same type of failure modes and instabilities obtained in practice, assuring that the analyses and criteria developed during this study will be tested against realistic events.

The work will be carried out in two phases. During the first phase, methods will be developed for an accurate recording of the progress of crack extension, and the crack size at instability. The possible use of NDI techniques such as ultrasonics and TBE enhanced flash radiography will be investigated. Techniques for measuring elastic-plastic strain distribution and crack opening displacement concurrent with loading and stable crack growth will be evaluated and refined as necessary. During the second phase, systematic measurements will be carried out and closely coordinated with the finite element modeling.

Two toughness-scaled materials with otherwise different properties (e.g., strength level, strain hardening index, etc.) will be selected. Approximately 35 experiments will be performed on center cracked panels with varying crack lengths of these materials. The largest panels will have a width of approximately 25 inches, the

---

*The toughness of A533B can be scaled by testing at subambient temperatures.
**$K_{Ic} \approx 280$ ksi$\sqrt{in}$., $\sigma_Y \approx 70$ ksi and t (wall thickness) ≈ 8 in.

smallest panels will be 5 inches. During these experiments
(and in others of this program), the following items will be
measured: (i) load, (ii) slow crack growth, (iii) crack size
at instability, (iv) strain distribution, (v) COD at several
locations along the crack face, and (vi) compliance. A comparable number of tests will be made on part-through cracked panels of "toughness-scaled" material and of A533B steel.

Heat treatments and testing conditions which produce a standard relative toughness ratio, tentatively
$(K_{Ic}/\sigma_y)^2 1/t \approx 2$ (or $0.7\sqrt{in.} \lesssim K_{Ic}/\sigma_y \lesssim 1.4\sqrt{in.}$ for plate
thicknesses in the range 0.25 in. to 1.0 in.) will be estab-
lished for about five candidate "toughness-scaled" materials.
The selection will be made from among a number of standard
aluminum alloys in sheet and plate form, high strength
steels, and A533B. The LEFM toughness values will be
estimated with the help of J_{Ic} procedures. The plastic flow
properties in two principal directions will also be obtained.

Efforts will be made to refine and support the concept of toughness scaling by demonstrating systematic relations between the appearance or mode of crack extension
(proportion of shear to flat fracture), the relative toughness
ratio $(K_{Ic}/\sigma_y)^2 1/t$, and more sophisticated expressions involving the absolute resistance to full shear fracture, flat
fracture, yield stress level and strain hardening rate. These
generalizations will be drawn from a review of the literature
and from experiments on the influence of thickness to be
performed during the study. The results will be tested
against the SwRI A533B flat plate experiments.

Systematic studies of the influence of thickness
will be performed on compact and DCB specimens of "toughness-scaled" materials. These experiments will be analyzed with the finite element model, and will provide information on the thickness dependence of G_c, R_c, $COD_c(a)$,
$J_c(a)$ and $R(a)$.

THE GENERAL ELECTRIC PROGRAM

This portion of the program is broken into four
phases as described in the following sections.

Selection of Viable Criteria for Crack Behavior During Plastic Fracture

The objective of this phase is to identify possible criteria that may describe the behavior of cracks during the phases of initiation and growth, and, through selected tests and analyses, to choose one or more viable criteria for further detailed study. In this phase, a number of candidate criteria that describe the initiation and growth of cracks will be evaluated by critical analytical and experimental studies. These studies will be carried out on compact tension specimens of A533 and weld metal. The analyses will be made through the application of finite element codes. Two codes are to be used: a two-dimensional code capable of describing elastic-plastic deformation, and the three-dimensional ADINA computer code (7), developed by K. J. Bathe of the Massachusetts Institute of Technology. The necessary subroutines for describing the crack tip and for calculating the various candidate criteria will be inserted into these codes. The outcome of the critical study will consist of the selection of one or more viable criteria for crack initiation and growth, which will be evaluated in more detail.

The criteria to be evaluated include those listed in Tables 1 and 2. For the case of crack initiation, emphasis will be placed on the J-integral and on the crack opening displacement, whereas in the case of crack growth, particular emphasis will be placed on the work done in a process zone at the crack tip, and on the strain amplitude, defined by the H-singularity.

The singular strain field at the crack tip is proposed for a criterion controlling crack growth, and this idea will be developed and applied in this phase. The strains near the crack tip are represented by the expression

$$\varepsilon_{ij} = H \, f(r) \, \tilde{\varepsilon}_{ij} \, (\theta)$$

where $f(r)$ is some singular function of r, and $\tilde{\varepsilon}_{ij} \, (\theta)$ is a function of the angle θ. The magnitude of the singularity is denoted by H; hence, H_{1g} becomes a logical criterion for crack growth, just as J_{1c} and K_{1c} are criteria for crack initiation. The implied condition of an attainment of a critical strain with the H_{1g} criteria automatically makes H_{1g} a function of plastic zone strength and size. Most crack growth studies suggest that the plastic zone ahead of the crack becomes increasingly larger as the crack grows. This information coupled with a H_{1g} criterion suggests that the applied load must be increased to sustain crack growth which is observed experimentally. Furthermore, the plastic zone strength and size (including the wake) are primary factors in the ultimate development of instability. Since the H_{1g} criterion is tied to these specific parameters, it may be a suitable parameter to detect the onset of instability.

To provide a common reference point for computation, the computer codes will be benchmarked through the solution of a sample problem (solved by assuming plane stress and plane strain). The same problem will be analyzed by Battelle investigators to provide a common benchmark of their two-dimensional computer code. In these benchmarks, only the situation before crack growth will be studied (i.e., nodes at the crack tip will not have to be released in the analysis). The crack growth situation will be analyzed subsequently as a continuing part of this project.

Several modifications are to be made to the existing computer programs. They include the ability to model the singular field at the tip of a crack (it is planned to use quarter point elements (8)), to evaluate the J-integral, crack opening profiles, and the work done in a crack tip process zone. The effect of the size of the finite element mesh on the accuracy of the results of a compact tension specimen will be determined. The optimum mesh size from a point of view of accuracy and cost will be used in subsequent analyses in the project. As the crack grows in a specimen, it is necessary to determine an analytically acceptable size of the increment of growth. Such a determination will be made based on numerical results of the finite element analysis.

The temperature that has been tentatively chosen for most of the testing is 93°C (200°F). This selection is done on the basis of the data in Figure 1, which shows the variation of strength ratio at maximum load as a function of temperature for A533 and NiCrMoV steels. The upper shelf behavior is reached near 93°C (200°F), and it is expected that tests at this temperature should simulate behavior at pressure vessel operating temperatures in the order of 250 - 300°C (480 - 570°F).

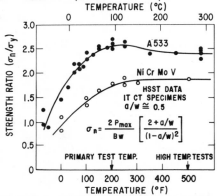

Fig. 1: Variation of Strength Ratio at Maximum Load with Temperature for A533 and NiCrMoV Steels (Source: Refs. 9 and 10)

For the finite element analyses, material property data will be obtained through tensile tests on A533, A508, and weld metal, at the appropriate temperatures. In order to check on the basic assumption that creep effects are negligible, relaxation tests at 290°C (550°F) will be made on specimens stressed to a plastic deformation state similar to that which might be experienced in practice.

The primary specimen to be used in the initial phases for experimentation will be the compact tension specimen. Each test configuration will be analyzed, and continuous interaction between testing and analysis will ensure proper interpretation of the results. All of the analyses at this stage will be two-dimensional. Initially, both plane strain and plane stress analyses of a compact tension specimen will be made in order to select the minimum specimen

thickness required for an essentially plane strain condition.
A key analytical task for evaluating crack criteria for initiation, growth, and instability involves the study of a compact tension specimen of A533 steel, having varying crack
lengths. The variation of the candidate criteria as a function of applied load or load point displacement will be determined. Critical values of the criteria for crack initiation
and crack instability will be determined based on the experimental observations. In parallel with the analysis of the
A533 base metal, compact tension specimens of weld metal
will also be tested and analyzed in order to ascertain the
validity of the candidate criteria for crack initiation and
crack growth in weld metal.

By testing 4T plane specimens, having thicknesses
from 25.4 mm (1.0 in.) to 101.6 mm (4.0 in.), an optimum
thickness of the specimen will be chosen for further testing.
The decisions on the smallest usable test specimens for
further testing will be based on measurements of the load-
displacement curves and on the calculation of J vs. δ a
resistance curves, as well as the observation of slow crack
growth. Using a single specimen thickness chosen on the
basis of the $93^\circ C$ ($200^\circ F$) tests, observations will be made
on the slow crack growth phenomena in two specimens at
$260^\circ C$ ($500^\circ F$) in order to ensure that the results are consistent with the $93^\circ C$ ($200^\circ F$) observations. If the results
are consistent, subsequent testing will be done at $93^\circ C$
($200^\circ F$). Because the COD is a candidate criterion for
crack behavior, an attempt will be made to develop a technique for its measurement, based on rubber infiltration
into the opened crack. The technique will be used at the
$93^\circ C$ ($200^\circ F$) testing temperature. Tests on about twelve
A533 compact tension specimens at $93^\circ C$ ($200^\circ F$) will be
conducted to obtain detailed information on load-displacement curves, critical values of J for crack initiation, crack
opening displacements (using clip gages and the rubber infiltration technique) and crack extension measured by the
compliance technique - target sensitivity 0.0254 mm (0.001
in.). Plastic deformation effects and metallographic features will also be examined. The results of these tests
will be used either as basic data for the finite element analysis to provide critical values of certain quantities (such

as J_c or COD_c) or as data to check the consistency of the conclusions derived from the analysis. Similar tests will be conducted at $93^{\circ}C$ ($200^{\circ}F$) on about six compact tension specimens of weld metal.

The influence of microstructure on the micromechanism of fracture will be studied with fractography and metallographic sectioning techniques. The mechanism of void nucleation will be investigated to show which inclusions, precipitates, or other microstructural features are void nucleation sites. It is planned to determine the levels and state of stress required for void nucleation by metallograhic sectioning of smooth and notched specimens which have deformed to several levels up to and including those required for fracture. The specimens used for this study will be analysed with elasto-plastic finite element calculations. The mechanism of crack initiation will be established by examining the highly strained regions at the crack tip in precracked specimens. Particular attention will be paid to the micromechanism of the void growth and coalescence processes which are required for macroscopic crack initiation.

The early stage of crack propagation will be examined utilizing compact tension specimens where the crack has initiated -- acoustic and electric potential techniques for crack growth detection will be explored. An attempt will be made to study the micromechanism of crack growth by observing the distribution of voids ahead of a crack which has experienced some stable growth. Metallographic observations of this type may also aid in the selection of the finite element node size and release rate of those nodes for crack growth modeling. These observations can also be used to experimentally measure crack tip opening displacement, crack opening angle, and the size of the crack tip process zone.

It is planned that the combination of these studies, and the results of the Battelle investigations, will lead to the next phase of the investigation concerning the development of a methodology for plastic fracture.

Development of a Methodology for Application to Flawed Structures in the Plastic State

The primary objective of this phase will be to make a final selection of the criteria that describe the initiation and growth of cracks in plastically deformed material. These criteria will be selected on the basis of additional tests on two specimen types fabricated of A533 steel: center cracked panels and double edge notched specimens. The analysis of these specimens will be based on two-dimensional plane strain finite element methods. After the selection of the criteria, additional tests will be made on a strip specimen to verify the criteria under plane stress conditions and on center cracked panels of weld material to verify the criteria in this material. Finally, three-dimensional analyses will be made of surface cracked specimens and compared with tests on similar specimens fabricated from A533 steel. As a result of these tests and analyses, criteria for crack initiation and growth will have been applied to a sufficiently wide variety of material and geometric conditions that a methodology for application to large scale structures will exist. The methodology will consist of criteria for crack growth and two- and three-dimensional analysis capability.

The choice of specimen types is dictated by a desire to attain varied states of hydrostatic stress and strain fields. The three specimens (compact tension, center cracked panel, and double edge notch) have considerably different plastic flow fields, as illustrated in Figure 2. The center cracked strip specimen has a deformation field that is constrained to the narrow strip of the specimen, giving substantially a different plastic flow than the center cracked panel. The former is chosen because its behavior is anticipated to be governed by plane stress conditions. The surface cracked specimen is chosen because it offers a three-dimensional plastic field. The specimens will be fabricated from blanks of size roughly 100 x 50 x 600 mm (4 x 2 x 24 in.). During this phase, roughly 35 specimens of A533B and 10 of weld metal will be tested at $93^{\circ}C$ $(200^{\circ}F)$.

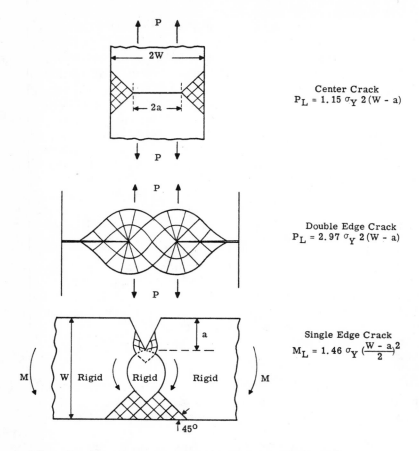

Center Crack
$$P_L = 1.15 \, \sigma_Y \, 2 \, (W - a)$$

Double Edge Crack
$$P_L = 2.97 \, \sigma_Y \, 2 \, (W - a)$$

Single Edge Crack
$$M_L = 1.46 \, \sigma_Y \left(\frac{W - a}{2}\right)^2$$

Fig. 2: Slip Line Fields for Three Specimen Types

During this phase of work, a decision will have been made, based on the accumulated analytical and experimental evidence, as to which criterion or criteria best describe the regimes of crack initiation, growth, and instability. In addition, it is expected that sufficient experience will have been accumulated in analysis and testing that at the end of this phase, a developed methodology for plastic fracture will exist.

Verification of Methodology Through Large Scale Tests

The objective of this phase will be to verify the criteria for crack initiation and growth in a large scale structural situation. Three methods of verification will be used:

● Large scale testing of flat plate specimens in bending. The plates will have holes with surface flaws situated at the corners. The strain distribution near these flawed areas will be chosen to resemble that in a pressure vessel nozzle corner. The material used will be A533 steel, and tests will be conducted at 93°C (200°F).

● Large tensile and bend specimens with embedded flaws will be tested at 93°C (200°F). These specimens will be fabricated from A533 steel and from weldment.

● The methodology for predicting plastic fracture will also be applied to the pressure vessel tests made during the Heavy Section Steel Technology (HSST) Program. In addition, the Battelle experience in the study of the SwRI flat plate results will be of value here.

Concerning first the tests on the flat plates, the proposed geometry will be a relatively wide and thick bend specimen containing a hole and a flaw as shown in Figure 3. The hole will be located on the tensile stress side of the

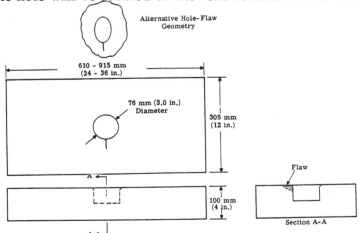

Fig. 3: Proposed Flat Plate Specimen

bend specimen, and a fatigue precracked flaw will be located at the corner region of the intersection between the hole and the specimen surface. This configuration simulates the main features associated with a flaw which may exist at the inside corner of a nozzle in a pressure vessel. Also, as indicated, an elliptical-shaped hole would be an alternative geometry. This would increase the elastic stress concentration factor in the region of the flaw. However, primary emphasis will be on the circular hole geometry since this produces an elastic stress concentration factor of approximately three, which is similar to that for the inside corner of a nozzle. The design will be analyzed using the three-dimensional ADINA code to assure a reasonable degree of plastic constraint. It is believed that a 76 mm (3 in.) diameter hole in a 305 mm (12 in.) wide specimen will have a sufficient degree of plastic constraint to be reasonably representative. However, if the analysis indicates more width is necessary, the dimension can be suitably modified.

The ADINA three-dimensional code, suitably augmented for calculating crack parameters, will be used, together with the chosen criteria for crack initiation and growth, to analyze the flat plate specimen with the surface crack present. The predictions will include the load-deflection curve of the specimen and the crack extension during loading.

About six flat plate specimens fabricated from A533 steel will be tested at 93°C (200°F). A large MTS servo-hydraulic testing machine will be used which has a load capability of 2.7 MN (600,000 lbs.). The bend specimens will be tested in either three or four points depending on which is more suitable after an analysis has been made. It may be possible to increase the specimen depth from the 100 mm (4 in.) indicated to 130 or 150 mm (5 or 6 in.), depending on the limit load behavior of the specimen. Of the six specimens, it is planned to test four with a circular hole and will have a variation in the initial crack size. The remaining two specimens will have elliptical holes. Measurements of the load deflection curve and the corresponding crack extension will be made. Crack initiation and extension will be measured using the compliance technique, as well as visual, ultrasonic, and acoustic emission monitoring techniques, where appropriate. Strain gage instrumentation will be employed. An existing compliance technique will have already been used to measure crack extension in the laboratory-scale specimens. Because of the much larger size of the specimens in this phase, this technique will be extended for the large-scale specimens. A target of 0.254 mm (0.01 in.) crack extension detection capability is expected.

Specimens containing a sharp tipped flaw totally embedded in the material are very difficult to produce. Figure 4 shows a specimen design which should simulate fairly closely the conditions around an embedded flaw. This specimen is fabricated from two half thickness pieces in which each piece contains a fatigue sharpened, semi-elliptical surface flaw. The two pieces are mated and welded along the edge with the weld root located fairly close to the edge of the flaw to provide a constraining effect. The welds will be stress relieved prior to testing. Embedded flaws in weldments will also be studied in this type of specimen by first making a weld in each half thickness piece, producing a flaw in the weld, and then mating and edge welding to produce a final specimen.

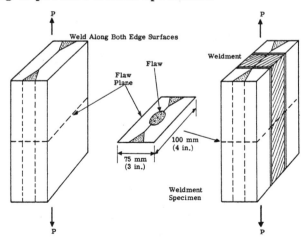

Fig. 4: Sketch of Candidate Embedded Flaw Tensile Test Specimen

The embedded flaw specimens will be tested as follows. All of the specimens will be instrumented with appropriate strain, extension, and displacement gages. Crack extension will be monitored by detailed visual and/ or ultrasonic techniques, and acoustic emission monitoring will be employed as appro riate. The feasibility of crack front marking by cycling at low loads will be studied. The tests will be carried out at 93°C (200°F) in tension. Similar specimens having offset embedded flaws will be also tested in bending. It is planned to test about eight of these specimens.

Application of Methodology to Representative Flawed Pressure Vessels

The objective of this phase is to apply the methodology and criteria for flaw behavior to representative nuclear pressure vessels. Application will be made to vessels undergoing hypothetical overpressure and thermal loadings.

Although rupture of any part of the reactor pressure vessels is never expected to occur, it is not possible to state with absolute certainty that it cannot happen under some hypothetical accident condition. Nevertheless, it has been shown by research carried out under the HSST Program that reactor pressure vessel materials can be expected to have sufficient toughness so that only with the simultaneous occurrence of very large flaws and very high stress is the rupture of the pressure vessel possible.

There are two identifiable hypothetical accident conditions which, at least in principal, can cause severe stresses to be induced in the reactor pressure vessel. These are the so-called LOCA (loss of coolant accident) thermal quench and the ATWS (anticipated transient without scram) overpressure. While these accidents are in the 10^{-5} to 10^{-7} probability of occurrence categories, they nevertheless are currently being debated in licensing and public hearings.

Consequently, the methodology developed during this study will be applied to the analysis of a representative nuclear pressure vessel, with specific attention being given to the LOCA quench and ATWS overpressure loadings. Recognizing that both of these loading conditions are system dependent, effort will be made to focus and generalize the fracture predictions relative to current design rules so that design margins can be determined.

The methodology will be applied first to surface and embedded longitudinal flaws in the body of the pressure vessel. Current analyses will be used to generate the stress boundary conditions necessary for applying the methodology. A series of flaw sizes will be inspected. Two-dimensional analysis will be applied in using the methodology. Flaw size vs. critical loading relationships will be generated for normal operating and overpressure (ATWS) conditions. The methodology will then be applied to nozzle flaws. Studies are planned on nozzle surface corner flaws and on nozzle weld flaws. One or two flaw sizes will be inspected. The three-dimensional augmented code will be used to analyze the critical load. The input for material and fracture properties of A533, weld metal, and A508 steels will be obtained from the results of the previous phases. Finally, the methodology for plastic fracture will be incorporated into current LOCA (transient temperature and thermal stress) analyses. Crack extension and instability will be predicted for longitudinal surface flaws and compared with Oak Ridge National Laboratory quench tests. Some additional studies will be made to assess the application of the methodology for light water reactor piping analyses.

SUMMARY

The planned program described in this paper contains elements of work oriented to both fundamental questions of crack behavior under large scale yielding, and to realistic applications concerning flawed pressure vessels. The program represents an amalgamation of a variety of disciplines, including mechanics, structural analysis and metallurgy. The program is designed to expose a variety of potential fracture criteria to rigorous experimental and analytical study. The two approaches by Battelle and General Electric are intended to be complementary, and to lead to the common goal of defining suitable criteria for application in the methodology of plastic fracture.

ACKNOWLEDGMENT

Many individuals have participated in the planning of this program, and will contribute to its execution. The authors wish to acknowledge in particular the assistance of D. Broek, M. F. Kanninen, A. R. Rosenfield, E. Rybicki, and F. A. Simonen of Battelle, and of W. Andrews, H. deLorenzi, D. Mowbray, T. Gerber, M. D. German, D. Lee, C. F. Shih, R. Van Stone, and S. Yukawa of the General Electric Company.

The program described in this paper is sponsored by the Electric Power Research Institute.

REFERENCES

(1) J. R. Rice, in Fracture, Vol. II, H. Liebowitz, ed., Academic Press, New York, 1968, pp. 191-311.

(2) J. A. Begley and J. D. Landes, in Fracture Toughness, Proc. of the 1971 Nat. Symp. on Fracture Mechanics, Pt. II, ASTM STP 514, American Society for Testing and Materials, 1972, pp. 1-23, pp. 24-39.

(3) A. A. Wells, Engineering Fracture Mechanics, Vol. 1, No. 3, 1969, pp. 399-410.

(4) P. D. Hilton and J. W. Hutchinson, Engineering Fracture Mechanics, Vol. 3, No. 4, 1971, pp. 435-451.

(5) J. R. Rice, "Elastic-Plastic Models for Stable Crack Growth," Mechanics and Mechanisms of Crack Growth, April 1973.

(6) S. C. Grigory et al., "Tests of 6-Inch Thick Flawed Tensile Specimens," HSST Program Reports HSST-TR-18, 20, 22, 23, and 24, SwRI, San Antonio, Texas 1972-1973.

(7) K.-J. Bathe, H. Ozdemur, and E. L. Wilson, "Static and Dynamic Geometric and Material Nonlinear Analysis," Report No. US SESM 74-4, Dept. of Civil Engineering, University of California, Berkeley, CA, 1974: also, K.-J. Bathe, "ADINA, A Finite Element Program for Automatic Incremental Nonlinear Analysis," Report No. 82448-1, Mechanical Engineering Department, Massachusetts Institute of Technology, Cambridge, Massachusetts, 1975.

(8) C. F. Shih, H. G. deLorenzi, and M. D. German, "Crack Extension Modeling with Singular Quadratic Isoparametric Elements," to appear in Int. J. Fracture.

(9) F. J. Witt and T. R. Mager, Nuclear Engineering and Design, Vol. 17, 1971, pp. 91-102.

(10) F. J. Witt, "HSST Program Semi-Annual Progress Report for Period Ending February 1972," ORNL-4816, Oak Ridge National Laboratory, Oak Ridge, Tennessee, October 1972, p. 46.

(11) J. G. Merkle, G. D. Whitman, R. H. Bryan, "An Evaluation of the HSST Program Intermediate Pressure Vessel Tests in Terms of Light-Water Reactor Pressure Vessel Safety," ORNL-TM-5090, Oak Ridge National Laboratory, Oak Ridge, Tennessee, November 1975.

APPLICATION OF NONDESTRUCTIVE EVALUATION IN THE AEROSPACE

INDUSTRY - RELIABILITY AND COST IMPACT

Ward D. Rummel

Staff Engineer

Martin Marietta Aerospace

Denver, Colorado

ABSTRACT

Advanced structural requirements and good engineering design practices emphasize a need for improved structural efficiency. Materials and design efficiencies have been emphasized in the aerospace industry to maximize the payload to carrier vehicle ratio. Development and application of linear elastic fracture mechanics as a design tool offers the potential for considerable improvement in materials efficiencies and, in some cases, substitution of lower cost materials. The advantages of fracture control principles can be realized only by reliable detection and characterization of flaws component materials. Research and development of nondestructive evaluation techniques in the aerospace industry has been redirected from detecting smaller flaws to reliable detection and characterization of flaws in functional hardware.

In addition to materials efficiency, emphasis on longer hardware lifetimes has been imposed to cut field maintenance costs. This paper discusses recent developments that implement fracture control in aerospace hardware. Recent work in establishing and demonstrating NDE reliability are also discussed.

INTRODUCTION

Nondestructive materials evaluation has been traditionally used in three basic functions:

(1) As an analytical tool for research development and engineering,

(2) As a process monitor/control tool for production and,

(3) As an acceptance tool in production and field maintenance.

In such use, the reliability of NDE has been unstated and/or assumed to be 100% in all applications. The results of such assumptions have been undisciplined and uncontrolled variance in inspection and process control reliability. Nondestructive evaluation and structural hardware management has been based on experience and engineering judgement and has varied with time and with engineering organizations.

In many cases, notes on engineering drawings have left the criteria for acceptance (i.e. certification for service) to the inspector performing the task. An example which is frequently encountered is:

"Inspect per Mil Std - 453A - No Cracks or,
No Defects or,
No Harmful Defects".

Evolution and acceptance of linear elastic fracture mechanics as a design/analysis tool has provided a common basis for development of acceptance criteria and for judgement of the effectiveness of criteria applied.

STRUCTURES DESIGN, ACCEPTANCE, RELIABILITY AND COST

The United States Air Force has incorporated technology developments in new procurements by specification of fracture control design requirements.

Procurement philosophy has been changed from a "safe life" approach to a "safe life/damage tolerance/fail safe" approach.

Essential elements which have been incorporated into this approach are:

. Damage tolerance in design
  - multiple load paths
  - slow crack growth.

. Initial <u>quantitative</u> certification of structural integrity.

. Service Maintenance based on
  - periodic quantitative recertification of structural integrity
  - maintenance schedule logic based on analysis of service environment, usage and predicted material behavior.

The costs associated with implementing the new philosophy have been analyzed and compared to costs which are incurred by using the previous philosophy. One area of major cost has been in maintenance of aircraft. In an assessment of problems during maintenance of aircraft, (1) the following were revealed.

1. Problem causes were due to:

   . fatigue and stress corrosion
   . few problems were due to overload and none were attributed to strength variations.

2. The source of the problems were identified as:

   . 60% due to pre-existing conditions,
   . 20% were service induced, and
   . 20% were design induced.

As a result of these assessments, emphasis has been directed to pre-existing conditions, i.e. conditions present at the time of aircraft manufacture. The cost of inspections performed on the production line is lower than the cost of set up and inspection at the maintenance depot. The effectiveness of inspections performed on new parts during manufacture is better than for those performed at the maintenance depot.

The Air Force Materials Laboratory has taken a commanding role in implementing the new procurement philosophy in the form of three documents:

. Mil Std-1530, How to Design
. Mil A - 83444, Damage Tolerance Requirements
. Mil I - 6870, Inspection Requirements.

The following abstracts summarize the essential elements of these documents which impact nondestructive evaluation technology application:

1. Use damage tolerance concepts in design analysis - assume the presence of pre-existing flaws.
2. Use basic materials fracture tolerance data in design.
3. Establish criteria for identifying fracture critical components.
4. Establish a fracture and fatigue critical parts list.
5. Zone fracture and fatigue critical parts on drawing with location of critical areas.
6. Identify flaw acceptance limits - size, location and orientation.
7. Identify inspection procedures to be used to reveal critical flaws.
8. If the critical flaw size identified is less than the assumed NDE capability, a capability to reveal critical flaws shall be demonstrated.

ASSUMED NDE CAPABILITIES		
Condition	Flaw Size (inches)	Material Thickness (inches)
Hole	0.050 thru the thickness.	0.050 or less
Hole	0.050 radius corner flaw	greater than 0.050
Other	0.250 thru the thickness	0.125 or less
Other	Semicircular flaw 0.250 (2c) length and .125 (a) in depth.	greater than 0.125

The National Aeronautics and Space Administration has
implemented similar requirements for the Space Shuttle program
in specific procurement documents and in:

> SP-8040 "Fracture Control of Metallic Pressure
> Vessels"
> SP-8057 "Structural Design Criteria Applicable
> to a Space Shuttle"
> SP-8095 "Preliminary Criteria for the Fracture
> Control of Space Shuttle Structures"

In addition to the cost advantages (i.e. life cycle cost)
which can be gained by implementation of the multiple dis-
cipline-structures acceptation philosophy, a balanced quantitative structures reliability program can be realized. When
fully implemented, the source of confidence in establishing
structures performance reliability can be varied. For example,
the primary source confidence may be:

> . design,
> . production, or
> . field maintenance.

For long life structures emphasis on design and production
can result in the most economical life cycle cost. For short
life structures, emphasis on production inspection and test
may be the most economical approach.

TECHNOLOGY CHALLENGES TO NDE

Challenges in structures design acceptance and reliability
analyses include major challenges to NDE engineering technology.
The philosophy must change from "How small a flaw can be detected? to "How large a flaw can be missed?" Reliable and
quantitative flaw detection, flaw size description and flaw
location/orientation must be developed and applied. NDE
reliability demonstration, NDE process control and NDE personnel qualification and periodic verification must be included. Several programs are currently in progress to meet
NDE challenges. These include:

> . "Recommended Practice for Demonstration of NDE
> Reliability" - The American Society for Nondestructive
> Testing.

. "Recommended Practice for Personnel Qualification-
SNT-TC-1A" - The American Society for Nondestructive
Testing.

. "Nondestructive Testing Standards Program" - U.S.
Bureau of Standards.

Several programs have been completed to establish and
demonstrate NDE reliability. (Ref. 2-11). From these data
files have been established and are under current analysis.
(Ref. 12, 13). Although the amount of validated NDE reliability data is limited at present, analysis of available data
can be used to evaluate current state of the art and to indicate areas where additional work is needed.

REVIEW OF AVAILABLE NDE RELIABILITY DATA

Martin Marietta Aerospace developed one of the most complete sets of data for inspection reliability on 2219-T87
aluminum alloy sheet.(5) Methods used to analyze these data have
been incorporated in the ASNT-Recommended Practice for Demonstration of NDE Reliability. In brief, 118 2219-T87 aluminum
alloy panels containing 328 fatigue cracks of varying sizes
were evaluated independently by three operators using X-
radiographic, liquid penetrant, ultrasonic and eddy current
techniques. This resulted in 984 observations for each
technique. Data were analyzed and plotted by starting at the
largest crack size, counting down 60 observations, calculating
the point estimate of detection probability (successes divided
by 60 opportunities) for the 60 observations, and plotting the
point estimate at the largest crack size in the sample range.
The process was repeated by sequentially counting down from
the longest remaining crack using the same sampling size. A
sample size of 60 observations is required to provide a 95%
confidence level for the data plot. Data are plotted at the
largest crack size to conservatively bias the data toward the
larger crack size. Figure 1 is an example of data plotted by
this method. The point estimate of probability is plotted as
a (0). The lower confidence limit for each data sample was
also calculated and is plotted as a (-). The area of primary
interest on this curve is the point at which the detection
drops below the 95% probability level. This is termed the
threshold detection limit and describes the largest flaw that
would be missed (at the 95% probability and 95% confidence

level) by the inspection technique. Various other plotting schemes have been used to plot data and have advantages for some purposes. (2) (12) (13)

The shape of curve is of primary interest in judging the validity of the data set. Typical curves representing three data sets are shown in Figure 2.

For controlled processes, the shape of the detection probability curve is known to take the form shown by Plot A. For randomly variable processes that approach control, the detection probability curves take the form shown by Plot B. For random uncontrolled processes that approach control, the detection probability curves take the form shown by Plot C.

Uncontrolled processes may be due to process variations or to application of a technique which is not effective for detection of the flaw under test.

Data plotted are for positive detection of flaws that are present. False reporting (Type 2 error) is not accounted for by this method. On cursory analysis, false reporting should not be of concern to the reliability of the hardware if detection is at the desired level. In practical tests of inspectors for X-radiography and liquid penetrant techniques, a high incidence of false reporting was found for inspectors who also missed test flaws. Conversely, inspectors who detected all flaws present, also had a low incidence of false reports. (14)

CONCLUSIONS:

NDE reliability and quantitative NDE present a broad challenge to NDE technology. Quantitative data and analyses are required to meet challenges of modern engineering methods. We have made a start and must generate the additional data and understanding to meet our challenges.

Ultrasonic Method, Set No. 3

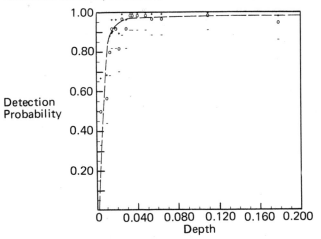

Figure 1.   Fatigue Crack Detection Plotted at 95% Confidence.
(Ultrasonic Method, Post Proof).

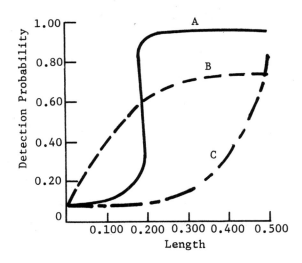

Figure 2.   Fatigue Crack Detection by Three Methods Plotted
at 95% Confidence

REFERENCES

(1)   Tiffany, C. F., "Forum on Capabilities of NDE Methods",
       Second Annual ASM Materials/Design Forum on Prevention
       of Structural Failure The Role of Quantitative Non-
       destructive Evaluation, April, 1974.

(2)   Packman, P. F., Pearson, H. S., Owens, J. S., and
       Marchese, G. B, The Applicability of a Fracture Mechanics
       Nondestructive Testing Design Criterion. TR-68-32. AFML,
       May, 1968.

(3)   Pettit, Donald E. and Hoeppner, David W.: Fatigue Flaw
       Growth and NDT Evaluation for Preventing Through-Cracks
       in Spacecraft Tankage Structures. NASA CR-128560,
       September 25, 1972.

(4)   Anderson, R. T., DeLacy, T. J., and Stewart, R. C.:
       Detection of Fatigue Cracks by Nondestructive Testing
       Methods. NASA CR-128946, March 1973.

(5)   Rummel, Ward D., Todd, Paul H. Jr., Frecska, Sandor A.
       and Rathke, Richard A.: The Detection of Fatigue Cracks
       by Nondestructive Testing Methods. NASA CR-2369,
       February 1974.

(6)   Bishop, C. R., Nondestructive Evaluation of Fatigue
       Cracks, Rockwell International - Space Division,
       SD73-SH-0219, September 1973.

(7)   Southworth, H. L., Steele, N. W., and Torelli, P. P.
       Practical Sensitivity Limits of Production Nondestructive
       Testing Methods in Aluminum and Steel, TR-74-241, AFML,
       November 1974.

(8)   Moysis, J. A. Jr., "Reliability of Airframe Instpections
       at the Depot Maintenance Level," Paper presented to the
       Spring Conferences, ASNT, 1973.

(9)   Caustin, E. L., B-1 USAF/Rockwell International NDI
       Demonstration Program, Rockwell International B-1 Division,
       1972-1973.

(10) Rummel, Ward D., Rathke, R. A., Todd, P. H. Jr., and Mullen, S. J., The Detection of Tightly Closed Flaws by Nondestructive Testing (NDT) Methods, NTIS-N-76-14475, October 1975.

(11) "Detection of Tightly Closed Flaws by Nondestructive Testing (NDT) Methods in Steel and Titanium, Contract NAS-9-14653, In Progress.

(12) Yee, B. G. W., et al, Assessment of NDE Reliability Data, NASA CR-50001, October 1975.

(13) "Methods for the Determination of the Sensitivity of NDE Techniques," Contract F 33615-76-C-5066, In Progress.

(14) Crockett, Lee, Rockwell International - Space Division, private communication.

NDT IN THE FLEET[+]

H. H. J. Vanderveldt*
S. Friedman**
*Naval Sea Systems Command, Washington, D. C.
**David W. Taylor Naval Ship R&D Center, Annapolis, Md.

INTRODUCTION

Nondestructive testing is part of the warp and woof of
any program of structural integrity assurance. The U. S.
Fleet, being the possessor and operator of millions of
tons of floating or submersible structure performs extensive
nondestructive testing (NDT) to achieve this assurance as an
integral part of corrective maintenance and periodic pre-
ventive maintenance. In what follows, we will give a brief
description of the tests performed and their purpose, the
acceptance standards and criteria, the conditions requiring
their performance, and the performing organization and
personnel. This description is not intended to be all
inclusive but is confined to the hull, propulsion (non-
nuclear), and auxiliary subsystems of the vessel. This will
serve to indicate how NDT fits into the Fleet's program of
structural integrity assurance and provide the frame of
reference for the summary of recent developmental efforts in
NDT that are in response to newer materials or structures or
increased component criticality. Much NDT is performed
during ship's fabrication prior to its acceptance by the
Navy for Fleet deployment. It will not be covered herein
directly, but many of the same techniques and criteria apply.

[+]The opinions or assertions made in this paper are those of
the authors and are not to be construed as official or
reflecting the views of the Department of the Navy or the
Naval service at large.

NDT TECHNIQUES

The procedures used in the various tests that are performed are called out in the documents listed in Table 1.
MIL-STD-271 covers four techniques: radiography, ultrasonic
pulse echo, magnetic particle, and liquid penetrant. The
NAVSHIPS' documents cover special applications of some of
these. NAVSHIPS 0900-006-3010 is applicable to ultrasonic
pulse echo inspection of hull welds, while NAVSHIPS 0900-001-
7000 deals with the ultrasonic pulse echo inspection of
brazed pipe joints. A brief description of the purpose of
each test and the means of assurance of test quality follows.

Table 1. U. S. Navy NDT Standards.

Designation	Title
MIL-STD-271	Nondestructive Testing Requirements for Metals
NAVSHIPS 0900-006-3010	Ultrasonic Inspection Procedure and Acceptance Standards for Hull Structure Production and Repair Welds
NAVSHIPS 0900-001-7000	Fabrication and Inspection of Brazed Piping Systems

Radiography - This technique is depicted schematically
in Figure 1.

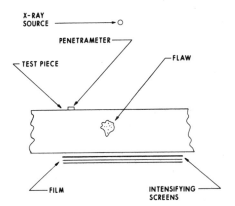

Fig. 1. Schematic Representation
of Radiographic NDT

The test piece is interposed between the radiation source and a film sensitive to radiation exposure. Upon subsequent development the film density, as measured by the percentage of light it transmits, will vary over its surface. Consequently, radiography is most sensitive to the presence of undesired voids and inclusions that result in a detectable film density variation insofar as they will transmit a different quantity of radiation than the surrounding sound structure. It is least sensitive to tight cracks with planes perpendicular to the direction of radiation and to inclusions of radiographically similar material. The quality is controlled extensively by the demands of MIL-STD-271, but the essential indicator is the specified penetrameter which is a thin shim of material (usually 2% of the thickness of the subject), similar to the test piece placed on it in a sort of worst case position prior to exposure. If the outline of the penetrameter and some of its holes is seen on the resultant radiograph, then the smallest hole seen attests to the resolution under conditions of minimal contrast. The outline of a penetrameter is shown in Figure 2, which depicts a seg-

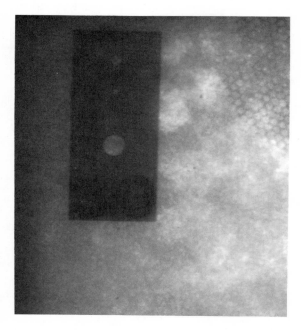

Fig. 2. Preliminary Reference Radiograph

ment of one of the many preliminary reference standards
prepared by the U. S. Navy in cooperation with the NDT
community. These efforts have culminated in the release of
documents such as ASTM 390, Reference Radiographs for Steel
Fusion Welds, a copy of which is shown in Figure 3. They
serve to aid the inspector in his identification of various
flaw types and severities.

Fig. 3.   Reference Radiograph Compendium

Ultrasonic Testing – This technique is most sensitive
to laminar flaws or tight cracks normal to the direction of
sound propagation although it will detect other types as
well. As is illustrated in Figure 4, periodic transmission
of bursts of high frequency (1-10 MHz) sound into the test
subject results in its reflection from any discontinuities
in its path. Reflections returning to the transducer may be
detected and displayed by the same electronics package that
was used to generate them. A typical 'A' scan shown in
Figure 4 is illustrative of the indications seen when
examining a flat plate with a cylindrical flaw. Both
longitudinal (compressional) and transverse (shear) waves

are employed, shear wave inspection permitting examination
of regions that would not be accessible to compression waves.

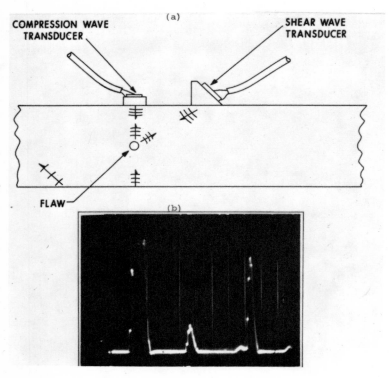

Fig. 4. Schematic Representation
of Ultrasonic NDT and A-Scan Display
Showing Flaw and Far Surface Echoes

The instrumentation calibration procedures are fully called
out in the standards cited previously. They yield assurance
that a minimal sensitivity and resolution are attained by
demanding initial performance of the system on specially
made, artifically flawed calibration blocks of composition
and surface finish equal to the specimen under examination.
For example, in the case of procedures for full penetration
butt welds called out in NAVSHIPS 0900-006-3010, we start
with a calibration standard depicted in Figure 5.

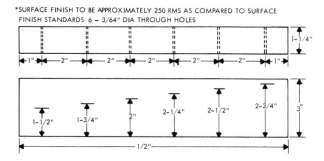

Fig. 5. Typical Reference Calibration
Standard (Sensitivity)

For either type of wave, instrument gain is adjusted until
the reflection from each and every one of the 3/64-inch-
diameter side drilled holes yields the same pulse height display amplitude, 20% of full screen. This is accomplished
by suitable adjustment of the required distance-amplitude
correction circuit. The gain is then increased by 12 dB.
The 20% line is then marked off as the disregard level (DRL);
while the 80% line is the amplitude rejection level (ARL) as
shown in Figure 6. Any signal below the DRL is ignored.
Signals greater than the ARL are usually automatic cause for
nonacceptance, while signals between the two must be subjected to further analysis. Linearity of response and
resolution of the instrument are also assured by other
procedures delineated in NAVSHIPS 0900-006-3010. Scanning
techniques and procedures receive extensive coverage in the
standards in order to assure completeness. For example, the
procedures prescribed for scanning of butt welds for longitudinal cracks may be described with the aid of Figure 7.
As is seen, the shear wave transducer is oriented so as to
direct the beam towards the weld at right angles to the
longitudinal axis. The combination of translational and
oscillatory motion of the transducer indicated in Figure 7
assures that the sonic beam sweeps the whole region of
interest in its search for longitudinal cracks or other
defects.

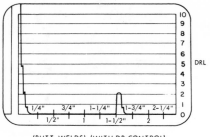

(BUTT WELDS) (WITH DB CONTROL)

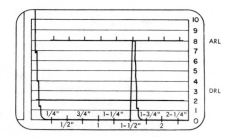

(BUTT WELDS) (WITHOUT DB CONTROL)

Fig. 6.   Viewing Screen Calibration

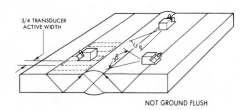

NOT GROUND FLUSH

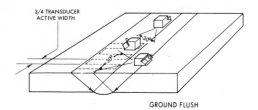

GROUND FLUSH

Fig. 7.   Scanning Procedure for Welds

Magnetic Particle Testing - Although this technique is
applicable only to ferromagnetic materials that will sustain
a high degree of remanent magnetization or have a high relative magnetic permeability, it is of great use to the Fleet
because ships' hulls are built of steel. The technique is
particularly useful when tight, linear subsurface flaws, to
which both radiography and ultrasonics are insensitive, are
suspected. As is seen in Figure 8, it consists of magnetizing
the test piece to saturation by application of a yoke or a
pair of current carrying prods. If a crack or other magnetic
discontinuity is near the surface and a component of the
applied field is normal to the crack plane then the resultant
dipoles at the discontinuity surface are the source of a
highly localized field near the discontinuity and exterior
to the surface.

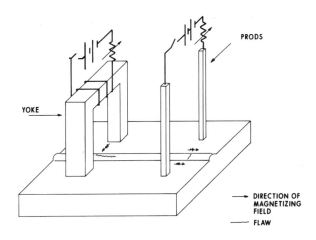

Fig. 8. Schematic Representation
of Magnetic Particle NDT

Upon application of the field, a finely divided ferromagnetic
powder is spread on the surface of the piece under test and
then gently blown away. Flux leakage at a discontinuity
magnetizes the powder causing it to be attracted to the surface of the test piece and remain behind after the rest of
the powder has been blown (or sometimes washed) away. Procedures for the conduct of an effective test are delineated
in MIL-STD-271. For example, the prod currents required
under various conditions of spacing and specimen thickness

by MIL-STD-271 are summarized in Table 2. The adequacy of
the magnetizing current or field is established by the use
of simple empirical formulae, or by insertion of a special
metallic indicator with artifical, subsurface flaws such as
the one depicted in Figure 9.

Table 2. Magnetic Particle Testing
Prod Current Specifications

Prod Spacing, Inches	Section Thickness	
	Under 3/4 Inch	Over 3/4 Inch
2 to 4	200-300 Amps	300-400 Amps
Greater Than 4 to Less than 6	300-400 Amps	400-600 Amps
6 to 8	400-600 Amps	600-800 Amps

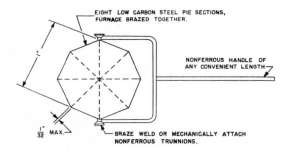

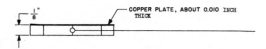

Fig. 9. Magnetic Field Indicator

Liquid Penetrant Testing - This method is applicable only to cracks or other void type defects that are open to the surface. The essence of the method as spelled out in MIL-STD-271 is outlined in Table 3. Processes such as surface preparation by grit blasting are avoided, since they tend to result in obstruction of the crack opening. A minimum penetrant dwell time assures full absorption by the defect. The major precaution in excess penetrant removal is the avoidance of removal of penetrant from the defect. The final step is the application of a developer, which is a finely divided white powder that will, in time, draw the penetrant out of the defect by further capillary action. The surface is then inspected under visible or ultraviolet illumination, whichever is appropriate, and indications are duly marked and recorded.

Table 3. Liquid Penetrant Inspection Procedure

- Prepare Surface
- Apply Penetrant
- Remove Excess Penetrant
- Develop
- Inspect
- Clean Surface

Eddy Current Testing - This technique has found its way into Fleet useage in specialized applications such as weld toe inspection and inspection of internally accessible copper/nickel tubing. It is covered by documents such as NAVSHIPS 0905-474-3010, the Standard Operating Instructions for Probolog Inspections. It is anticipated the basic technique will be covered by the next revision of MIL-STD-271. The basis for the technique is illustrated in Figure 10. The impedance of a helically wound conductor is affected by being in close proximity to a metallic or other electrically conductive body by interaction with the field of the eddy currents induced in the conductor by the field of the helical coil. Changes in electrical conductivity, inductor to metal spacing or electrical discontinuities affecting eddy current flow will all result in coil impedance variations. Thus,

eddy current inspection can be used to ascertain thickness
of coatings on metallic surfaces, determine the extent of
corrosion of copper-nickel tubing, and the detection of
cracks along seams covered by paint or otherwise unsuited
for visual inspection. The standards used to calibrate the
instrument are usually artifacts that more or less simulate
the condition or defect of interest. Most instruments are
designed to be adjusted to ignore impedance changes due to
nonrelevant factors such as the variation of spacing between
the inductor and the surface (i.e. lift-off) when looking
for cracks in seams.

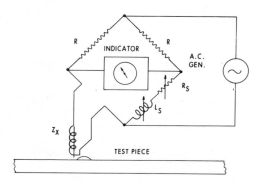

Fig. 10. Schematic Representation
of Eddy Current NDT

SHIPS SUBSYSTEMS TESTED

The subsystems that are at times subjected to the tests
that have just been described are listed in Table 4 along
with the specifications and standards that determine the
conditions under which NDT is to be performed and point
towards the acceptance criteria to be found in detailed
description in still other documents. There are three basic
situations under which NDT is performed: new construction,
repair, and certification of structural integrity. The first
two are covered by the documents we have listed, while certification is confined to higher criticality situations beyond
the scope of our presentation. By and large, NDT is performed
on a repair in the same manner and to the same acceptance
criteria as for the original structural element. The decision
to make the repair will arise from some scheduled certification or unscheduled inspection process.

Table 4. U. S. Navy NDT Performance Requirements

Component	Designation	Title
Hull	NAVSHIPS 0900-000-1000	Fabrication, Welding, and Inspection of Ship Hulls
Submarine Hull	NAVSHIPS 0900-006-9010	Fabrication, Welding, and Inspection of HY-80/100 Submarine Hulls
Machinery, Pressure Vessels, Piping	MIL-STD-278 (SHIPS)	Fabrication, Welding, and Inspection; and Casting Inspection and Repair for Machinery, Piping, and Pressure Vessels in Ships of the U. S. Navy
Piping (Brazed)	NAVSHIPS 0900-001-7000	Fabrication and Inspection of Brazed Piping Systems

ACCEPTANCE STANDARDS

As may have been gathered from our preceding remarks, these are not a statement of what is acceptable under a specific set of circumstances but rather the establishment of various levels of quality indication, one of which may then be set down as a requirement for a particular structural element. These various quality levels are defined in the documents listed in Table 5. Starting with the radiographic standards, NAVSHIPS 0900-003-9000, we come across the use of terminology employed for all NDT performed in accordance with MIL-STD-271: namely, Class 1, Class 2, and Class 3 as designations for levels of quality. Class 1 acceptance criteria permit fewer or less extensive defects than Class 2 which, in turn, is more stringent than Class 3. Indications of incomplete fusion or penetration, slag, porosity, and undercut are all specified in terms of both the size of an individual indication or the number of indications per linear inch of weld. Generally, the thicker the weld, the more the extent of the allowed defect for a given class. Thus, one large indication or many small indications per unit length may both be the basis for rejection. By way of example, Figure 11, reproduced from NAVSHIPS 0900-003-9000, graphically defines the acceptance criteria for both individual and cumulative incomplete fusion and penetration indications for Class 2 and Class 3 welds. Turning now to

the ultrasonic criteria set down in NAVSHIPS 0900-006-3010,
we find the same Class 1, 2, and 3 designation as in radiography.

Table 5. U. S. Navy NDT Acceptance Standards

Technique	Designation	Title
Radiography	NAVSHIPS 0900-003-9000	Radiographic Standards for Production and Repair Welds
Ultrasonics	NAVSHIPS 0900-006-3010	Ultrasonic Inspection Procedure and Acceptance Standards for Hull Structure Production and Repair Welds
Liquid Penetrant, Magnetic Particle	NAVSHIPS 0900-003-8000	Surface Inspection Acceptance Standards for Metals

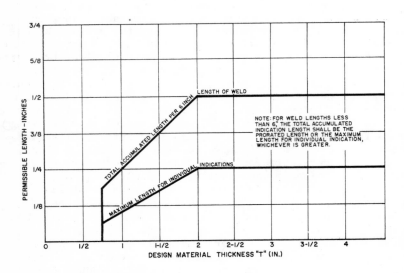

Fig. 11. Class 1 Production Weld Acceptance
Standard for Incomplete Fusion and Incomplete
Penetration Standards

The severity of an indication is given both in terms of the
amplitude of the observed signal and the apparent length of
the defect producing the indication. Signals that exceed
the established amplitude rejection level (ARL) are automatically unacceptable for the most stringent Class 1 criteria.
Signals of less than the disregard level (DRL) are ignored,
while signals between the ARL and the DRL will only be the
basis for nonacceptance if the indicated discontinuity
exceeds a specified length that is a function of weld thickness and is significantly greater than that indicated by
echoes greater than the ARL. The criteria for acceptability
are at times rather complex, but necessarily so. For example,
in the evaluation of cumulative, adjacent discontinuity indications less than the ARL, a scheme such as shown in Figure
12 applies. Two discontinuities are seen, the ends being
separated less than twice the greater discontinuity length.

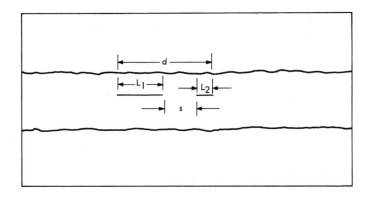

TOP VIEW

CODE

L = LENGTH OF DISCONTINUITY

s = MAXIMUM SPACING BETWEEN ADJACENT
    DISCONTINUITIES

d = MAXIMUM DISTANCE BETWEEN
    OUTER EXTREMITIES

DATA

L_1 = 3/4 INCH

L_2 = 1/4 INCH

s = 1/2" (LESS THAN $2L_1$)

EVALUATION IS BASED ON D OR

$L_1 + L_2 + s$

Fig. 12. Evaluation of Adjacent Discontinuities

Consequently, the effective discontinuity length is 1½ inches, which would be cause for rejection of a weld with a thinner member of less than 3 inches thickness in the case of a Class 1 weld. In the case of the penetrant and magnetic particle criteria delineated in NAVSHIPS 0900-003-8000, the same class designations still apply. Indications are of course visual under either visible or ultraviolet illumination and acceptance in a given class is based on both the length of an individual indication being less than a certain thickness dependent value, while the accumulated length of indications smaller than this individual length must again be less than a certain value per unit length in a manner dependent on the specimen thickness and the length of the individual indications. Again, some necessary complexity prevails in the specification of cumulative lengths of discontinuities.

An example of the acceptance standards for cumulative small, nonlinear indications may be described with the aid of Figure 13.

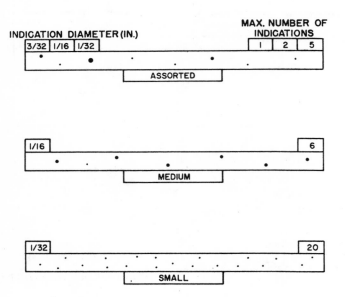

Fig. 13. Acceptance Standard for Cumulative, Nonlinear Indications

The basic criterion for acceptance is that the total indication area for the length shown shall not exceed 0.75% of the weld surface area. The illustrations show various combinations of indication diameters that just pass. If the actual indications depart significantly from the examples given, then, in effect, the diameter of each indication must be measured and the total indication area computed on the basis of the indication being a circle.

TEST PERSONNEL AND ORGANIZATION

At this point, it is apparent that the successful performance and interpretation of the test is very dependent on personnel skill and knowledge. MIL-STD-271 requires that all testing be done by personnel qualified to criteria established by the American Society for Nondestructive Testing, Publication ASNT-TCIA. There are three levels of qualification: Operator, ASNT Level I, performs the test; Inspector, ASNT Level II, interprets the test results and accepts or rejects the part; and Examiner, ASNT Level III, has the experience and sagacity that befit one who is authorized to train, examine, and certify the qualification of a potential operator or inspector. In accordance with Chapter 9922 of the Naval Ships Technical Manual, U. S. Navy enlisted personnel may, if meeting certain prerequisites, be trained and certified at the Service Command School, San Diego, California. The general requirements for qualification to each level are set down in this document for all of the techniques covered by MIL-STD-271 and for visual inspection as well. One must qualify separately for each technique. After training, the candidate is examined on the basic theory, interpretation of specifications, and a performance of an actual test. He is then assigned to one of the three types of ships' tenders where he provides the first echelon of NDT service. Their higher echelon civilian counterparts at Naval bases or shipyards must meet the same essential certification requirements. All are subject to regular or ad hoc requalification in each of the techniques. If all this sounds rather elaborate, one might recall the old adage that the least reliable but the most important part of an automobile is the nut behind the wheel. So to, in ultrasonics its the nut appended to the transducer that's the most important part of the system.

NEW TECHNIQUES

The formidable NDT arsenal brought forth to do battle against the forces of structural unreliability in the U. S. Fleet is still wanting for improvement based on the following: current procedures are much too dependent on operator skill. In addition, they are tedious and present safety hazards in the form of dangerous X or gamma radiation. Over and above this, new materials and structures are vying for a place in the future Fleet, and we must be ready for them. Air cushion vehicles and hydrofoils will be much more prone to fatigue than conventional craft, and they will use high strength to density ratio materials and joints that are not always amenable to our classic NDT approach. Table 6 summarizes the current Navy program and thinking in Fleet-required NDT developments.

Table 6. Naval Sea Systems Command
Program in NDT; Exemplary Elements

- Acoustic Emission Monitoring for Structural Integrity Assurance

- Ultrasonic Data Processing and Display in Real Time

- Acceptance Criteria for Thin Section Aluminum Weldments

- Rapid, Accurate Measurement of Residual Stress

The first program element encompases a further Navy development and evaluation of the acoustic emission phenomenon. It is based on the fact that crack initiation or crack growth in a metal will generate stress waves with the initiation or growth site as their source. By reception and analysis of these stress waves, crack initiation can be detected and localized and growth of particular cracks may be monitored. This will serve to supplement and also eliminate much Fleet NDT. Currently, it is necessary to periodically examine for new cracks as well as have a look at the old ones to see if they have grown. In between, we worry about them. With an acoustic emission monitoring system that does its intended job, we will eliminate the worries and confine our NDT attentions to newly reported or newly enlarged cracks. A developmental version of this system is shown in Figure 14.

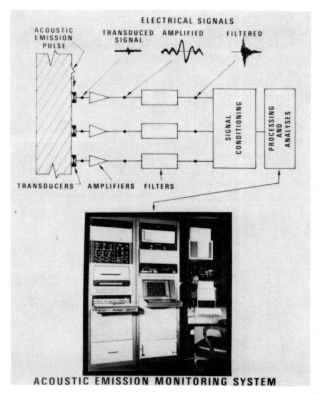

Fig. 14. Acoustic Emission Monitoring System

The next element is part of a larger program concerned with the development of an automatic weld tracker for ultrasonic testing of welds. An earlier version of this is shown in Figure 15. The real-time data processing and display will, even under manual scanning under Fleet conditions, eliminate the current tedium of ultrasonic data recording by the operator and provide both a real-time and hard copy display of defect location and geometry for evaluation and for future reference. This will also help eliminate the need for radiography in many instances.

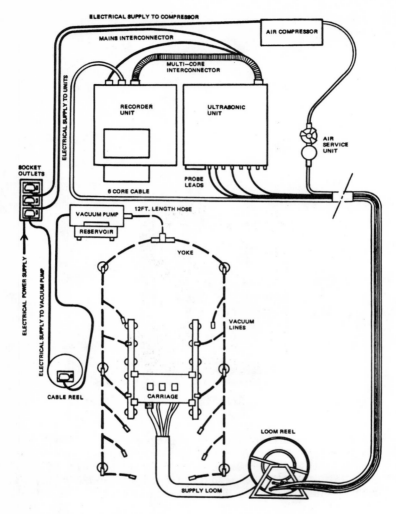

Fig. 15. Early Version of Automatic Weld Scan Tracker

The next element, dealing with the acceptance criteria
for thin section aluminum weldments, is motivated by the
current trends in hydrofoil and hovercraft design. A
typical ship of this type is shown in Figure 16. Therein,
welded structures of aluminum alloys are extensively used
or proposed. A review of current specifications and

standards reveals that thin section aluminum weldments are
not adequately covered. Over and above this, there is the
technique problem of detection of slag by radiography,
alumina being almost equivalent radiographically to aluminum,
and the replacement of magnetic particle testing by an
equivalent technique suited for nonferrous materials.

Fig. 16. Example of a High Performance
Ship; the Hydrofoil

The last program element points to a relatively new
approach to NDT and to structural integrity. Although
residual stress measurement methods are well known, accurate
methods are prohibitively time consuming if a survey of
large extent is involved. On the other hand, more rapid
measurements must be investigated further before they can be
used for Fleet survey of critical structural areas and elements. If found suitable they will be used to establish and
monitor areas or structures more prone to fatigue or corrosion
fatigue failure.

CONCLUDING REMARKS

With these last remarks we bring our brief presentation
to a close. We have given you a bird's eye view of how NDT
is done in the U. S. Fleet and what developments we may
expect in the future.

346 / H. H. J. Vanderveldt, S. Friedman

We would like to acknowledge the valuable assistance of
Mssrs. J. Gleim of the Naval Ship Engineering Center and
S. D. Hart of the Naval Research Laboratory in providing many
of the facts and perspectives contained herein.